Science Through Stories

Teaching Primary Science with Storytelling

Chris Smith PhD & Jules Pottle

Foreword by Pie Corbett

Storytelling Schools Series
Volume III

Hawthorn Press

Hawthorn Press

Published by Hawthorn Press, Hawthorn House,
1 Lansdown Lane, Stroud, Gloucestershire, GL5 1BJ, UK
Tel: (01453) 757040 Fax: (01453) 751138

E-mail: info@hawthornpress.com

Website: www.hawthornpress.com

Illustrations by Karen Donelly and Shirin Adl

Design and typesetting by Lucy Guenot

Printed by Berforts Information Press Ltd, Herts, UK
Reprinted 2017, 2018, 2019 by Henry Ling, The Dorset Press, Dorchester

Printed on environmentally friendly chlorine-free paper manufactured from renewable forest stock.

British Library Cataloguing in Publication Data applied for.

ISBN 978-1-907359-45-3

About the authors

Chris Smith PhD

Chris Smith, PhD, is a storyteller, educational trainer and the founding Director of Storytelling Schools. Chris loves helping to make education more joyful, effective and engaging, especially in areas of social deprivation where good education can make such a difference to future life chances. For the last ten years Chris has been researching and developing the Storytelling Schools idea in UK schools. Chris has also been a father, musician, exhibition designer, performer, monk, UN manager, human ecologist, surfer and writer. He currently divides his time between a house in Oxford and a wood in Devon. For more information on Chris see: www.storysmith.co.uk

Jules Pottle

Jules Pottle lives in Oxford with her family and has been working in primary schools for over twenty years. She is passionate about teaching primary science in an exciting and engaging way. She believes in plenty of talk and practical work and loves writing science stories to use in the classroom. She has been a class teacher and science coordinator for many years and is currently working as a science specialist teacher in the local primary school where she developed the stories for this book. She is also a freelance museum educator at the Story Museum, provides INSET training on primary science around England and is a trainer with Storytelling Schools. In her spare time, she enjoys putting on productions with young people and adults in her local community theatre group.

Storytelling Schools

Storytelling Schools is a group of educationalists who are passionate about the power of storytelling to transform education and learning. We were founded in 2013 to promote the Storytelling Schools model throughout the United Kingdom and beyond. By systematically learning storytelling skills at school, students can receive an education that builds confidence and fluency in spoken language, raises standards of reading and writing and provides an engaging way to learn other subjects in the curriculum, all in a method that is enjoyable and inclusive.

Storytelling Schools make information, resources and training available for teachers who wish to adopt this approach in their school. Our founders are Chris Smith, Adam Guillain, Pie Corbett and Nanette Stormont. For more information see: www.storytellingschools.com

Acknowledgements

First and foremost we must acknowledge Pie Corbett. Many of the ideas of the Storytelling School are based on Pie's remarkable and inspiring work. Chris had the pleasure of working and learning with Pie for several years as the Storytelling Schools idea developed. We appreciate his support, inspiration and brilliance. The approach also leans on learnings from two great teachers, storyteller and Crick Crack Club chief Ben Haggarty and the wonderful Alida Gersie, expert on so many aspects of storytelling and education, whose continued support is much appreciated.

Thanks also to Martin, Claire and Merry at Hawthorn Press for their support, understanding and patience, and to Anthony Nanson for his thorough, thoughtful editing which greatly improved our final manuscript. We are lucky to have such a great editor.

Thanks to the staff and students of Cumnor Primary School, where these stories and methods were first field tested, and to Jules's long-suffering family for their patience and support during the writing of this book.

Contents

Foreword by Pie Corbett viii

Preface x

Chapter 1
Introduction 1

Chapter 2
Introducing the Storytelling School Approach 9

Chapter 3
Biology Stories 21

Topic 3.1	SEEDS AND SEASONS	21
	Story: Anna's Apple Tree	22
Topic 3.2	SENSES	29
	Story: Going to Catch a Piggy-wig	30
Topic 3.3	HUMANS, HEALTHY DIETS, TEETH AND BONES	37
	Story: Uncle Jack	38
	Story: The Smiling Princes	42
Topic 3.4	HABITATS	49
	Story: Mummy, Can I Have a Penguin?	50
Topic 3.5	CARING FOR THE ENVIRONMENT	57
	Story: The Drop of Honey	58
	Story: The Bird and the Forest Fire	64
Topic 3.6	HEALTH AND GERMS	71
	Story: The Story of Edward Jenner	72
	Story: The Broad Street Pump	76
Topic 3.7	FLOWERING PLANTS	81
	Story: Jack and the Giant's Peach	82
	Story: Emily's Bees	88

Chapter 4

Chemistry Stories		**93**
Topic 4.1	USES OF MATERIALS	**93**
	Story: The Fairy Godmother's Day Off	**94**
Topic 4.2	CHANGING MATERIALS	**101**
	Story: Death of a Pancake	**102**
Topic 4.3	ROCKS AND FOSSILS	**107**
	Story: The Fossil Woman	**108**
Topic 4.4	PROPERTIES OF MATERIALS	**115**
	Story: The Horses of Troy	**116**
Topic 4.5	THE WATER CYCLE	**125**
	Story: The Children of the Water God	**126**
Topic 4.6	MIXING AND SEPARATING MATERIALS	**135**
	Story: Nimblefingers	**136**

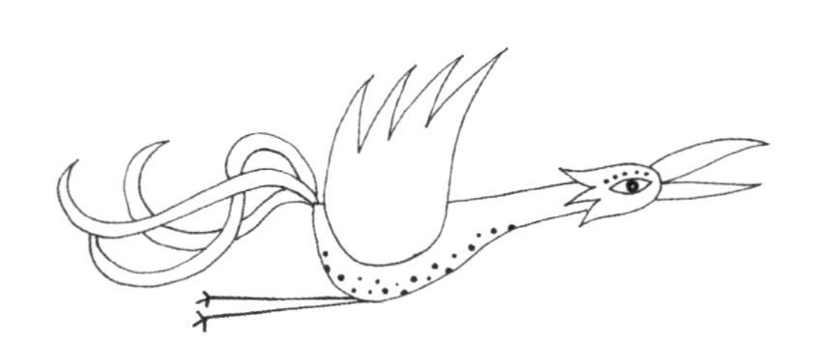

Chapter 5

Physics Stories

Physics Stories		143
Topic 5.1	FORCES – PULLING AND PUSHING	143
	Story: The Giant Turnip	144
Topic 5.2	LIGHT AND NIGHT AND DAY	149
	Story: The East and the West	150
Topic 5.3	SOUND AND HEARING	157
	Story: Little Rabbit Goes Home	158
	Story: The Blind Man and the Hunter	164
Topic 5.4	FORCES – PUSH, PULL AND TWIST	171
	Story: The Wheel that Jack Built	172
Topic 5.5	FORCES – MAGNETISM	179
	Story: The Magic Stone	180
Topic 5.6	LIGHT	187
	Story: The Torch	188
Topic 5.7	ELECTRICITY – SERIES CIRCUITS	197
	Story: The Lighthouse Keeper's Son	198
Topic 5.8	SOUND	207
	Story: The Bat Who Learned to Click	208
Topic 5.9	SPACE	213
	Story: Apollo 13	214
Topic 5.10	ELECTRICITY – PARALLEL CIRCUITS	221
	Story: The Rocket	222
Topic 5.11	FORCES AND BRIDGES	229
	Story: Bernie's Bridge	230

Sources and Notes 238

Index 243

Foreword

Our most memorable teachers use storytelling to bring their subjects alive. Story draws us in, makes us care about the subject and sparks the imagination. It has always been used by great communicators and should be a part of every teacher's repertoire.

This book explores the world of science through story. Some of the stories act as a springboard into scientific investigation or communicate information. Others are about great scientists and their discoveries – stories that every child should know.

When I look at the moon, I become a child again, revisiting that sense of awe I must have felt on first seeing the moon's eye. The fingernail curve of mercury seems to beckon. It is an image that echoes with mythology. Stories whisper in the darkness as the moon casts its silver light. But there is more than feeling and story to the moon. What is the moon? Why does it wax and wane? This is another mode of thinking, of interacting with the world.

Science feeds off our curiosity about the world, the desire to explore and find out about the delicate mechanism and intricate design of life. In a way, science is no less magical for being based in reality. You only have to watch David Attenborough sitting by a gorilla and hear his voice to know that science is about wonder as much as story can be. Science is also about play. Playing with ideas to explore, investigate and find out. Without the ability to create, to come up with fresh ideas that might be of use, our scientific understanding will wither. Story also lives in playfulness. Stories tease the imagination as we recreate them for ourselves. When we invent stories, we are playing. The alliterative twins story and science dwell in the realms of creativity, playfulness and wonder.

The eye of a bee seen under a microscope opens a new world of beautiful surprise that we never knew was there. We have to learn how to see, how to take time to discover the beauty of our surroundings, how to appreciate. Science opens up this world and sparks our curiosity. Whether we're watching hungry caterpillars munch the edge of a leaf or growing beanstalks for Jack to climb, our sense of wonder blossoms.

Children need time to pause at the side of the playground and watch as the spider scuttles along its web, to see how sunlight through the dewdrops on the web creates tiny rainbows; to lean on the windowsill at night and listen to the silence; to notice their hearts beating and pause to rub thyme between their fingers or smell the mint. The excitement of making a light for the cardboard lighthouse, of the moment when the wire touches the battery and the light flickers on, of staring at the filament as it glows – all of this is the magic of science. Most wonderful and awe-full of all is that moment when the ends of two magnets push against each other and we feel an invisible force. What is that strange resistance that cannot be seen? That is surely as magical as a minotaur pacing the darkness or a spider that speaks.

Stories are one form of magic. Words strung well together grow the imagination and inside us new worlds are created. Stories make us laugh, feel anger or a wash of sudden sadness. They are powerful tools for building ourselves and our culture. Science is another form of magic and it is one that is real and beneath our hands, in front of our eyes. Story is the inner magic. Science awakens the outer magic. To be aware of the intricate and delicate workings of the world we live in, to nurture curiosity and care for ourselves and the environment is as necessary to education as knowing the stories of our folk. Science savours the world and stories nourish our minds.

If there is the inner magic of story and the outer magic of science, then an education with either side of the scales missing is imbalanced and impoverished. We need to know our own story and be able to tell it, for that is who we are and without the telling of it we do not know ourselves and neither will anyone else. The stories of our people and world help to explain and sustain us through the passages of life and its great odyssey; but those stories alone are insufficient.

The story of the bee; the tale of the seeds as they become plants; the apple blossom that turns to fruit; the apple that contains an orchard; the cress that needs water and light; the miracle of water becoming ice; the struggle to support a bridge until the right structure is found – all of these things are part of the great story wheel of who we are, our world and its survival. Stories feed our imagination and the world is revealed to us through the imagination. We only have to learn to pause, and look and wonder at what is within and around us.

Without science we are lost. Without story we are trapped alone in the darkness of ourselves. For too long, these companions have wandered on separate tracks. This book takes one positive step forwards to bringing them together as travelling companions.

Pie Corbett

Preface

This book will show you how to teach science in primary school using storytelling as a launch pad for a topic. The collection of 29 stories is intended for *retelling*, first by the teacher and then by the students. Broadly, we use the 'Storytelling Schools' approach to teaching. This involves children telling stories from memory as a strategy for learning both language and topic content. It combines storytelling, drama and creative writing in a systematic way across the school and has roots in Pie Corbett's talk-for-writing system.

You can find out more about the general approach in Chapter 2 of this book and in greater detail in Volume 1 of this series – *The Storytelling School Handbook for Teachers* by Chris Smith and Adam Guillain – which emphasises how to use storytelling to teach language and raise standards in writing.

At the time of writing there are more than 100 Storytelling Schools in England. The approach is now a tried and tested way to teach both literacy and many cross-curricular topics in the primary curriculum.

In this volume, we show how these storytelling methods can be applied to the teaching of primary science, with a series of stories that can be used *both* as a springboard for science learning *and* in teaching literacy.

This book is written for all primary school teachers. You don't need to be a science specialist or to work in a Storytelling School to use it. If you are *not* teaching in a Storytelling School, you can still use the stories and ideas in this book to teach primary science, and if you are using a topic-based curriculum then you can use the stories to lead your teaching of science-related topics.

The teaching of science in primary school plays a key role in introducing every child to scientific ideas about the world and how it works. This first experience of science is likely to influence later choices about study and career.

Science is important in so many ways to our society: personally, practically, socially, economically and aesthetically. Scientific explanations and scientific enquiry are part of everyday life. Most things, whether living or not, whether gas, liquid or solid, can be understood using scientific ideas, from the subatomic (what it's made of) to the cosmic (where it comes from) and everything in between. Science is such a wonderful, vast and mysterious branch of human knowledge, which helps us to understand our own place in the universe. Our job in primary school is to communicate the excitement of science, so that children can engage with the awe and wonder of this hands-on subject.

Learning to tell stories with explicit science content offers a great way to make these first experiences memorable, enjoyable and inclusive, because most children naturally engage enthusiastically with hearing, retelling and exploring such stories. In this way, the science can be learned indirectly, building on the class's natural enthusiasm for story, so the stories provide a meaningful context for science teaching. As primary teachers, you have the chance to be a gatekeeper to that amazing world. Stories can help you provide that gateway.

Chapter 1

Introduction

Science education and storytelling

This book is for primary school educators and shows how to use storytelling to teach primary science. It offers a collection of bespoke stories that you can use as a springboard for teaching the most common primary science topics. Some have been created specially for science teaching, some are traditional tales and some are stories based on history or biography. They cover the main science topics taught in English primary schools. We have listed them in general order of difficulty rather than for a particular year group, since year groups vary so much and many of the stories can be adapted to a range of ages and abilities.

Main science subjects covered in this book:

Biology (in order of difficulty)

Topic Focus	Story	Other science topics related to the story
3.1 Seeds and seasons	Anna's Apple Tree	Reproduction, human and plant life cycles, requirements for life in plants
3.2 Senses	Going to Catch a Piggy-wig	Animals, food chains, predators
3.3 Healthy lifestyles	Uncle Jack The Smiling Princes	Humans, healthy diets, teeth and bones
3.4 Habitats	Mummy, Can I Have a Penguin?	Environment
3.5 Care for the environment	The Drop of Honey The Bird and the Forest Fire	Interrelationships, ecosystems, forests, fire, birds
3.6 Health and germs	The Story of Edward Jenner The Broad Street Pump	Scientific method, vaccination, sanitation, prevention of disease
3.7 Flowering plants	Jack and the Giant's Peach Emily's Bees	Life cycles, conservation, pesticides, habitat, ecosystems

Chemistry (in order of difficulty)

Topic Focus	Story	Other science topics related to the story
4.1 Uses of materials	Fairy Godmother's Day Off	Vehicles
4.2 Changing materials	Death of a Pancake	Cooking, butter making, metalworking, foxes
4.3 Rocks and fossils	The Fossil Woman	Dinosaurs and other extinct species
4.4 Properties of materials	The Horses of Troy	Greek machines
4.5 The water cycle	The Children of the Water God	Changing materials, changes in state
4.6 Mixing and separating materials	Nimblefingers	Polar bears

Each story comes with three things:

1. tips on how to tell and teach the story;
2. a range of ideas for ways of linking the story to the teaching of literacy and other curriculum areas;
3. a list of possible ways to teach the related science topic.

These ideas are offered as suggestions and are not meant to be prescriptive. You can mix and match according to the needs and preferences of your class. The stories themselves can likewise be adapted to fit age and ability.

The book is a collaboration between a primary science teacher and a storyteller and the topics have been chosen to support the National Curriculum in England. The stories and teaching ideas have been tried and tested in the classroom. We like the combination of engaging with the imagination and playfulness of storytelling and the experiential exploration provided by practical science investigation. Between the two, there is usually something for everyone. Some children are drawn to practical matters and enjoy explaining how things work. They may engage more enthusiastically with stories that include scientific problems and explanations, which thereby support their creative development as well. Other children love fiction, and so fictional stories enable them to explore the science in a way that feels familiar and interesting to them.

Great stories naturally inspire curiosity. Retelling the stories in various ways builds confidence and skill in communication of the story and the ideas within it, including the science concepts. At the same time, stories can provide a great way to develop curiosity about the science content, and confidence to explore and apply that content, first within the story and then beyond it.

Physics (in order of difficulty)

Topic Focus	**Story**	**Other science topics related to the story**
5.1 Forces: pull	The Giant Turnip	Food, plants, animals
5.2 Light and night and day	The East and the West	Planets, space, light as a requirement for plant growth
5.3 Sound and hearing	Little Rabbit Goes Home The Blind Man and the Hunter	Habitat
5.4 Forces: push, pull and twist	The Wheel that Jack Built	Machines, rivers
5.5 Forces: magnetism	The Magic Stone	Compasses and navigation
5.6 Light	The Torch	Electricity
5.7 Electricity: series circuits	The Lighthouse Keeper's Son	Lighthouses and navigation
5.8 Sound	The Bat who Learned to Click	Bats, habitat, evolution
5.9 Space	Apollo 13	Sun and moon, gravity, orbits
5.10 Electricity: parallel circuits	The Rocket	Space, space travel, gravity, orbits
5.11 Forces and bridges	Bernie's Bridge	Properties of materials, structures, buildings

How does this apply to primary science teaching? Primary science is all about two worlds: the world around us and the world within us (our own bodies and minds). It is all about investigating and explaining things we come into direct contact with every day.
For example:

- When we look up into a clear night sky, what do we see?
- When a switch turns on a light, what is going on?
- Why does my ice-lolly melt?
- How does a baby grow into a child and then an adult?
- Why do we get sick?
- How do my lungs work?

These kinds of questions may prompt investigation of our direct, everyday experience. Things that children see and hear and touch provide ripe material for exploring those two key questions: 'Who am I?' and 'What is the world around me like?'

A challenge for every primary teacher is how to bring these questions alive, so that the class is engaged and involved in finding satisfying answers. The students have to care about the questions and the possibility of an explanation. Some students are naturally keen on investigating and explaining the physical world, whereas others may not quite see the point: I may feel it is enough for me to know how to switch on the light. Who cares how the light works?

The thesis of this book is that one way to create that *caring* is through storytelling. If we care about what happens in the story, then we will be involved in both the characters and the content of the story. In this way, students become engaged in scientific learning.

Take the example of the light switch. One way to explain about electricity could be to explain about circuits and power sources, then have the students play around with batteries, wires, lights and switches, making all kinds of circuits, before they talk and write about their experience. That's one way to do it. Some students will learn well that way. However, others may not be so well engaged.

Now imagine starting the topic by telling a story about lighthouses (see Topic 5.7):

Once there was an island with three lighthouses manned by two brothers and their father. Every night they would light the oil lamps so ships would steer clear of the rocks, and every morning, as the sun rose, they would snuff out the lamps. The third and youngest son lived with his father. He always seemed to have his nose in a book.

'Why waste your life reading? What's the point of that?' his old father would say. 'Why can't you do something useful like your brothers?'

'Dad!' the boy answered. 'I am studying! This is useful.'

The father got so angry he threw the book on the fire.

Then one day the two older brothers were called up to fight in the war. Who would light and care for the lamps?

'Don't worry, Dad,' said the youngest son. 'I've got an idea. There's this new thing I read about called electricity. We can use it to light up all the lighthouses at the same time.'

'Lad, books won't help light the lamp,' the father said and he turned to go.

'Listen, Dad,' said his son. 'I've got an idea. If we buy a generator, wires and some electric lamps, we can turn all the lighthouses on at the same time with a single switch!'

The father looked doubtful ...

This story fragment provides a background of human drama and tension, in this case the father–son conflict, which draws us into the shoes of these two characters. The human interest and 'what-happens-nextness' of the narrative gives a meaningful context for what follows: a story about power supplies, circuits and the validation of the son's studying. The father's view of the son is transformed and the boy finds his place in the world. On the way, both the characters and the story's listeners learn all about power supplies, circuits, switches and lights.

Human minds are hard-wired for stories. We naturally allow our minds to be absorbed into a story once we have been hooked by it. Stories are among our most popular forms of relaxation and entertainment. So a narrative, well told, can draw the whole class into the story context, providing an inclusive springboard for further teaching and a relevant context for teaching the electricity topic. If the story is memorable and meaningful then the science will be memorable too: i.e. the story is the hook.

We have written these stories not just to be told by the teacher, but also to be retold by every member of the class. This is a great way of anchoring the language and ideas of the story in the memory of your students.

To do this, we recommend using the 'HMSS' method: hear, map, step, speak. After hearing and discussing the story, students can learn to retell the story in their own way by mapping it, 'stepping' it (freeze-framing the main story moments) and then speaking it in pairs and groups a few times in their own words, adding their own pieces of description and speech. In this way, students forge a deeper connection with the story. They will naturally embody the characters' actions, thoughts and feelings in order to be able to tell the story. To do this, they need to clearly imagine the events as they unfold. The scientific content of the story is then remembered and retold along with the rest of the tale. It becomes archived in the memory through its rehearsal and performance. This is a deeper embedding of memory than will be gained from simply reading or hearing the story once.

The task of retelling gives the students a new purpose: to tell the story in a satisfying way. This adds a new kind of caring enabling purposeful rehearsal and performance.

Once the story has been retold a few times, there are then many teaching options. For example:

- Drama and role-play can explore story elements including science explanations and description.
- Practical experiments can recreate the process of wiring up the lighthouses with wires, batteries and bulbs.
- Students may create a similar story involving burglar alarms or football floodlights.
- Students may role-play fantasy lectures by an electricity professor explaining what happened or how the systems work.
- Students may read around the subject to deepen their understanding of the story's context and enrich their telling of the story.

At any point in this process, the oral and practical work can be linked to a piece of writing, such as fiction or a report. In this way the student will learn science and literacy at the same time: one stone, two birds.

Science in every story

Scientific ideas can be applied to almost everything about ourselves and our world. So science content is always implicit in many elements of any story. Stories are full of things that can be used as starting points for teaching science: this means that, while this book gives you examples of stories to work from that have a specific science focus, you can also find science-learning opportunities in any story. Here's an example using a version of the ubiquitous *Little Red Riding Hood* (LRRH) and listing some of the science-learning opportunities that may be linked to the story.

Story section	**Possible science links**
In the kitchen LRRH learns from her mum that her grandmother is sick with flu, so she has to take eggs, milk and fruit so her grandmother can become healthy again.	• Infectious diseases • Eggs and chickens • Cows and milk • Healthy eating
She walks into the forest smelling the mulch and mould of the leafy soil. She sees squirrels, foxes, badgers blackbirds and buzzards. Around her, between the oaks, are a carpet of bluebells and clumps of ferns. The stones are damp and mossy. She hears birdsong from the tree canopy, and then ... the howling of a wolf. Her heart beats fast as she imagines the danger ahead.	• Forest ecosystem: plants, animals, their interactions and life cycles • The senses • The heart
Coming to a fork in the road, she sees a huge grey wolf staring down at her, teeth gleaming, tongue lolling, ears pricked up. 'He's going to eat me up!' she thinks, frozen to the spot with fear as she imagines becoming wolf food.	• Wolves • Fear (freeze or flight)
Just then she hears two woodcutters whistling behind her on the path. She turns to look back – and the wolf is gone. She breathes a sigh of relief and feels her muscles relax: the danger is over!	
She takes the long winding path and spends time in clearings warming her face in the sun, picking flowers and collecting nuts for her grandmother. When her basket is full, she goes back to the shade of the forest.	• Trees, flowers, nuts • Light and shadow
Eventually, she comes to the cottage and knocks. 'Come in,' says a voice; a strange, deep voice. In the gloom inside she can't see much, so she goes over to the bed. It is dark and damp and cold and she shivers. 'What big eyes you have.' 'All the better to see you with.' '... ears ... nose ... mouth you have ...' Then the wolf gobbles her up!	• Sense organs

Clearly, there are elements in some stories which are not realistic (talking wolves) and so contradict some aspects of scientific knowledge, but that need not be an obstacle to linking this story to learning about real wolves.

Whatever story you are working with in your class, there will always be opportunities to deepen the student's engagement with the story by exploring story elements from a scientific point of view. This can deepen curiosity and embed knowledge while anchoring the story more deeply in the student's imagination. It may be quick, e.g. Why was it cold in the house? Or it could be more involved, e.g. Let's do some wolf research! Where there is interest in the story already, you can build on that interest to teach new skills and knowledge.

After the story, what next?

Obviously, what's next depends on what you want to teach. For science teaching you may want to explore related topics in a practical way and devise writing exercises to go with them. For literacy teaching you may wish to link all this to narrative writing or non-fiction writing in various ways.

Generally, in a Storytelling School a typical teaching sequence might go like this:

- teacher tells the story;
- students learn to tell the story;
- deepening activities to anchor the story deeply in the mind and imagination;
- shared writing linked to narrative writing of the story (see p. 17);
- creation of innovated versions of the story for retelling and writing;
- creation of non-fiction pieces of oral communication and writing which are related to the story.

To put it another way, once the story has been learned, there is first a period of deepening in which the story and its content are explored through research, drama, role-play, art, poetry. After that, the story may be written using the learned plot, or innovated versions may be created for subsequent telling and writing. The teaching of the standard non-fiction communication types (persuasion, discussion, recount, explanation, instruction and information reports) can also be linked to the story theme and its science content. Non-fiction pieces may be performed orally (presentations) and/or developed into written pieces (see p.17).

If you'd like to know more then read on to Chapter 2, where we spell out the Storytelling School approach in more detail, or better still get hold of a copy of *The Storytelling School Handbook for Teachers* (by Chris Smith and Adam Guillain) for a more thorough treatment of the approach and its origins and applications.

This process of deepening, innovating and writing can be directly geared to the teaching of science. For example role-play and drama provide a good way to refine and develop science explanations orally, while practical hands-on experiments can deepen the understanding of the science in the story. Writing exercises linked to the science can further develop competence in scientific communication. All this fits easily within the Storytelling Schools scheme.

What's in the rest of the book?

The next chapter looks at Storytelling School ideas in more detail. The rest of the book contains the stories and with each story some suggestions for ways of working with the story once it has been learned using HMSS.

Each story has been classified by plot type and story genre, since these are often taught explicitly in Storytelling Schools. For plot types we use the seven basic plots, from Christopher Booker's *The Seven Basic Plots* (Continuum, London, 2004). The seven plots are:

1. voyage and return
2. rags to riches
3. overcoming the monster
4. quest
5. tragedy
6. comedy
7. rebirth

For genre we use a variation of story classification developed from the 'ladder to the moon' scheme as taught by the storyteller Ben Haggarty. The genres include:

- historical fiction (based on historical events)
- general fiction (new literary story)
- folktale (realistic content, traditional style)
- fable (includes speaking animals and other creatures, clear moral)
- wondertale (includes magical creatures)
- legend (told as past true story with fictional elements)
- myth (includes gods)
- creation myth (involves creation by gods and forces in nature)

For more details of the use of plot types and genres in Storytelling Schools, see the chapter on 'Invention' in *The Storytelling School Handbook for Teachers* (pp. 61–76).

We have divided the stories into biology, physics and chemistry topics. All stories have been written in a storyteller's voice to make it easier for them to be learned and retold by the teacher and the class. They have been kept short to make them easy to read and retain. Each story comes with:

- story and topics summary;
- tips for retelling;
- ideas for deepening, including deepening the science focus of the story;
- possible ways to link to literacy and other topic teaching;
- other ideas for teaching the science content of the story.

These are not lesson plans, but rather examples of things you might include and adapt in your planning according to the needs of your class. The stories may be used for any primary class, but you may need to adapt and adjust the telling and sequence of activities to suit the age group and ability. The stories are linked to the topics in the National Curriculum in England, which does allocate particular topics to each year group, so you will find that the stories linked to easier concepts topics are aimed at younger children but you can always tell a simple story to older children and adapt the activities to fit the class. Likewise, you may want to simplify a story to tell it to younger listeners and simplify the activities accordingly.

Please try to retell the stories yourself. Of course, they can be read to the class, but take a little time to learn the story using HMSS (see p. 14) and you will engage and motivate the class much more fully and provide a role model for their own retelling. Great teaching usually includes storytelling: these tales are a good way to do it. The first few stories may take a while to learn, but after three or four, most people can learn and retell a story in a few minutes: a great skill for any teacher to have in their teaching toolkit! Remember, you don't have to recite the stories word for word: the main method involves remembering the main events and then finding your own words to narrate them: much quicker and easier than word-for-word reciting and much more engaging for your audience.

Chapter 2

Introducing the Storytelling School Approach

The basic idea is simple. Storytelling and storymaking are a great way to develop communication skills, social skills and topic knowledge all at the same time.

In a Storytelling School children learn to tell and improvise stories from memory as a way of learning these things in a systematic way.

These different areas of learning are often linked together in the following sequence over the course of a few weeks:

- First, the teacher tells the story.
- Next, the children learn to tell the story themselves, supported by map-making and story-stepping.
- After that children engage in activities to deepen the story in their imagination, developing ideas and new language to include in their retellings.
- Then the teacher may conduct a shared writing class to teach the writing of the story before allowing the children to have a go themselves.
- Alternatively, the children may create innovated stories with new settings, characters or activities based around the original story.
- After learning to tell these new stories, they may then create new written versions.
- In addition, children will sometimes invent completely fresh stories and learn to tell them before creating written versions.
- Reading may occur at the deepening stage, when children get to know setting and characters, and at the writing stage, when they may study sample texts of various kinds.
- The same kind of teaching sequence may then be applied to non-fiction pieces, first learning orally and deepening before linking to writing and creating innovated versions.
- The content of the story may also be linked to the teaching of various subjects across the curriculum.

The Storytelling School elements are often linked together as shown in the diagram below.

How the Storytelling School elements fit together

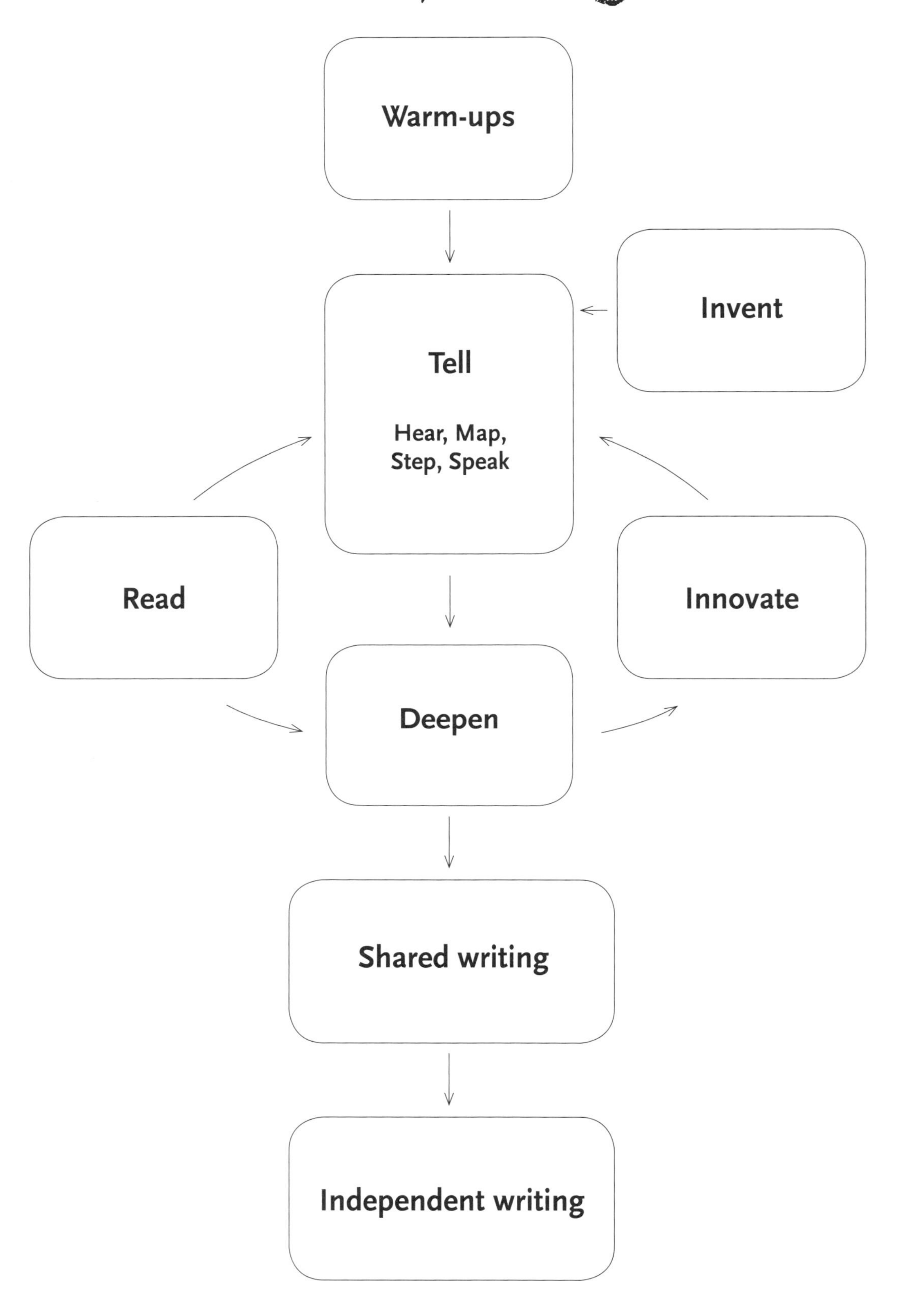

The rest of this chapter will explore the following common questions:

- What kinds of stories are told and retold?
- Are they remembered or improvised?
- How does this help with language development?
- How does storytelling relate to writing?
- How do students remember and retell stories?
- What about non-fiction?

What kinds of stories are told?

Usually Storytelling Schools start with 'traditional' stories from oral tradition. These are the easiest and often the most fun to tell, since they have evolved for this purpose. However, any kind of story can be told orally, though some are easier than others! Once the storytelling method has been mastered, children in Storytelling Schools may tell literary stories, history stories, religious stories, personal stories and of course science stories as well as their own fantasy stories. The same method can also be applied to non-fiction presentations of various kinds.

We have found it useful to distinguish three ways of performing a story. All three can be used by teachers to tell a story and then by children to retell.

1. **choral stories**, which are chanted together and so learned word for word;
2. **participative stories**, where the audience joins in in various ways;
3. **independent stories**, which are told by one teller while the audience listens without actively participating in any other way.

All three kinds of storytelling can be useful for science teaching. Teachers can choose stories and the method of performance depending on the aptitudes and learning needs of the class or individual concerned. Participative telling, allowing constant physical and oral engagement in a flexible way, is the most popular way of telling for 4–7-year-olds. Choral storytelling is more common in early years but can be a good way for primary children to learn things word for word (for example poems and scripts). For the 8–11 age group independent telling becomes more suitable as students develop the capacity to listen easily without needing to join in. Most of the stories in this book are intended for participative or independent telling.

Are stories told word for word or improvised?

Both are possible. Choral stories may be learned word for word, whereas in participative and independent telling children are taught to remember the main events and 'pictures' of the story and then to tell the tale in natural way using language that naturally arises as the story is being spoken.

Although some adults may be daunted by the idea of improvising a story, most discover that it is much easier to remember main events than all the words. Most remember a sequence of pictures and gestures, which are enough for them to retell the story with fluency. This process also teaches skills of sequencing, condensing and concision, all of which are valuable for many forms of learning and communication. A story can often be captured in a map of 10 or so pictures. Each picture references hundreds or even thousands of words that might be used to describe that part of the story. Instead of remembering all the words, you just have to remember 10 pictures: it's much quicker and easier than memorising thousands of words.

How does this help with mental development?

Language development and thinking are intimately connected, because we think many things through and understand them using language. By learning complex language we learn complex thinking.

Besides accelerating language learning, storytelling helps hugely with sequencing skills, which are crucial to skills of memory, reflection, recounting and presentations. Storytelling also helps children learn to plot patterns of various kinds, so they understand how to make their own stories by reusing the patterns of a plot. This develops the skills of pattern discernment, which is great for budding writers (and also budding scientists)! It is also a valuable way of rehearsing innovation, which may then be applied to any kind of creative or original thought as well as being a great springboard for writing. Start them off early and they will go far!

How does learning stories relate to making up stories?

Our colleague and mentor Pie Corbett came up with this brilliant way of thinking about how learning stories relates to making them up. The basic idea is this: when it comes to making up stories it's really a matter of recycling. If a child knows lots of stories already then they will have a store of ideas to draw upon when they make up their own. They will have ideas for characters, settings, problems and solutions and will weave them into their stories. Conversely, a child who has no knowledge of stories will struggle to make up their own, or will simply draw on personal experience. If you want to produce great storymakers then fill them up with amazing stories.

Pie has described the process of linking storytelling and storymaking as comprising three stages:

1. **imitation**, where the child learns to retell a story they know, keeping the story more or less the same;

2. **innovation**, where the child changes the story to some extent with new settings, characters, events or resolutions;

3. **invention**, where the child makes up a fresh story without basing it on a known story.

Consider, for example, *The Billy Goats Gruff.*

At the imitation stage the child will stick to the learned plot:

Three goats cross the bridge one at a time. The first two trick the troll and the third butts him off into the next valley. Then the goats go off to the hill and feast on long green grass.

The child will tell this story either word for word or improvising pieces of speech and description but not changing substantially what happens.

At the innovation stage the child begins to change the story. The goats may become pigs, the troll may become a wolf, the bridge may be in a city and the green grass may become a burger bar. A child may innovate a bit, let's say changing the goats to hamsters but otherwise keeping the story the same. Or the child may innovate a lot, retelling the story as a journey of three friends down a scary alley haunted by a witch who wants to eat them as they go to the sweet shop. All this is the realm of innovation. It helps the child understand the plot pattern and reuse the language patterns in interesting and engaging ways by which the child can feel pride and pleasure in their new story.

They can also change the plot in other ways, deleting parts (for example fewer goats) or adding things (a goat ninja), but substitution is the most popular form of innovation.

At a higher level still, the child may simply recycle the idea. *The Billy Goats Gruff* is a story about overcoming a monster. They might make up new stories about monsters in which the monster is overcome in other ways than that in the original story. By this stage the child is beginning to invent independently of the original story.

More independent invention comes when the child simply makes up a story without reference to any previous tale. There may be a theme or an idea or a prompt but then the child is free to make the story up any way they wish.

Here, the child will recycle the stories they know into the new tale. In a Storytelling School, the child is provided with tools to help them do this in a satisfying way. The plot matrix is one such tool; it provides prompts for creating stories that work well. At the beginning, the story usually needs a setting, a character and a core problem or dilemma (first three boxes); the problems are not immediately overcome and often get worse before getting better. A helper may arrive and become involved. Finally, at the end, problems are solved (or not), the story finds an ending and resolutions (or not) and there is chance to reflect on learning. This matrix can be used to create, analyse and improvise stories, highlighting the main elements that will need development. A great tool for both teachers and students.

1. Where	2. Who	3. What (problem)
4. Obstacle	5. Setback	6. Helper
7. Solution	8. Ending	9. Learning

How does storytelling relate to oral communication skills?

In a Storytelling School children learn satisfying ways to communicate with one another. Stories are learned so that they can be retold to others. This gives purpose to the process of learning the story. Children discover the satisfaction of finding their own voice and knowing it to be enjoyed and appreciated. They learn that sustained listening can be a pleasure. Building such confidence is a crucial part of primary education, not least because, in a digitalising world, personal one-to-one communication seems to be on the wane. Often, when the Storytelling School approach is begun in a school, staff remark that the social and emotional maturity of children improves dramatically. This is not only because children learn to work together, but also because when we tell stories we have to empathise with the characters for the story to work well. In this way the storytelling approach not only builds vocabulary and language skills but also increases the capacity to understand and relate to others.

How does storytelling relate to science teaching?

By making a link between something already familiar and a new idea, stories can provide us with a way to explain something new to ourselves so that we can picture it and understand it. Some of our stories explain scientific concepts by giving scientific processes a narrative and characters. For example, the story about the water cycle gives the process of evaporation a character, with a face and hands and a reason to make the visible liquid water into an invisible gas. The sequence of the story follows the sequence of the water cycle, so remembering the narrative helps to make the sequence of the cycle more memorable.

When we begin to understand a new concept, most of us will construct a visual image for ourselves about what is going on. When you listen to a story, rather than watch a film, you use the same skill, constructing a visual image of what is happening in the story. Some of the stories in this book have been specifically written with this in mind. In order to emphasise that a complete circuit – a loop – is needed for electricity to flow, the lighthouse story contains a loop of lighthouses on an island that all go off when the wire is broken. The image of the circuit on the island being in a loop is there to help the children to visualise the circuit and retain their learning.

Telling and retelling the stories also help the children to practise explaining the science in the story, recalling the facts and events in the correct sequence and thereby internalising the science. The act of telling gives us a purpose to keep revisiting these sequences and become more fluent and familiar with the science in the process.

Excellent primary science teaching includes a good deal of talking. Children need to ask questions in order to communicate what they would like to find out. They need to explain how to set out an experiment, how to make it fair and what they expect to happen. They need to describe what they have observed and they need to explain the science behind their results. Lastly, they need to analyse their own experiments in order to improve their experiment design. With all this explanation happening in the classroom, it makes perfect sense to explicitly teach explanation formats (spoken and written) alongside the science. Spoken rehearsal can then be used to help them learn to write non-fiction texts. After each of the stories in this book, there are lots of suggestions for activities that will give children opportunities to practise non-fiction text types orally.

Excellent science teaching also engages the children in hands-on activities in which they can experience the science they are studying. We have included plenty of easy science activities to go with each story. You can choose which activities you include in your planning. Although non-fiction writing is an important part of science teaching, it should only comprise about a third of the lesson time. For this reason, we suggest you arrange the recording of their work so that they only do one part of it in each lesson. In one session, you could ask them to write a really precise prediction, in another they could write the instructions and in another they could explain what they found out. There is no need to record every part of an experiment, every time.

How do students learn to remember and retell stories?

Our favourite way of learning to tell and improvise a story employs a neat four-step process called HMSS, or hear, map, step, speak. HMSS was developed from techniques taught by the great storyteller Ben Haggarty. It works like this:

HEAR. First, the children hear the teacher tell the story.

MAP. Then, the teacher and/or children create a story map of the main 'marker' events in the story. A visual summary is made, linked to the audio and imagined memory.

STEP. Next the children 'step' the story, finding a word or sentence and gesture for the key marker moments in the narrative. These may or may not be exactly the same as the marker events. The idea is they take a step forward between these moments, like walking over stepping-stones, and find a movement and phrase for each stone.

SPEAK. Once the 'bones' of the story have been learned through mapping and stepping, the child is ready to start telling the story in their own words, sometimes in pairs, sometimes in a larger group. The telling starts to put 'flesh' back on the bones of the story, using the child's own ideas while sticking to the same plot.

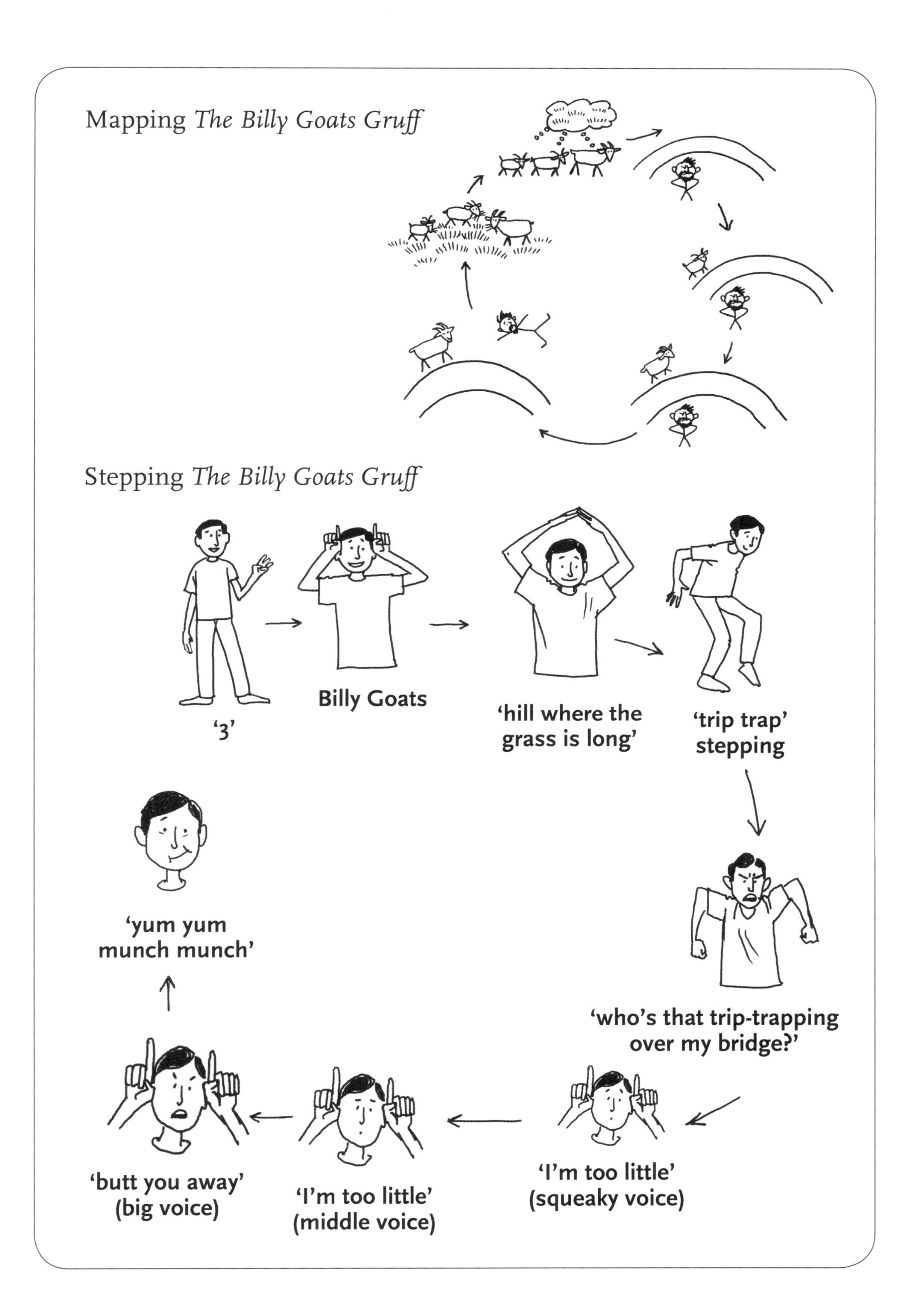

Mapping *The Billy Goats Gruff*
Stepping *The Billy Goats Gruff*
'3'
Billy Goats
'hill where the grass is long'
'trip trap' stepping
'who's that trip-trapping over my bridge?'
'I'm too little' (squeaky voice)
'I'm too little' (middle voice)
'butt you away' (big voice)
'yum yum munch munch'

HMSS is fast and effective because it mixes various learning models in one clearly defined and easily learned process. You can use it to remember any sequence, so it can come in handy for all kinds of learning.

What do we do next once the children know how to tell the story?

Often the next step is what we call 'deepening', when children develop their own ideas for developing the story. This may include:

- whole class re-enactment;
- hot-seating;
- role-playing;
- inventing gestures and sounds and words for the story;
- inventing songs and chants;
- producing ideas for speech and description;
- drawing pictures of parts of the story or its characters, with annotations;
- interview role-play on a TV show.

All these allow the story to develop in the children's imagination in a playful way.

How does storytelling relate to writing?

The principle here is obvious: the stories fill the child up with words and ideas, which can then be translated into writing. Because the child already has lots of things they want to communicate, they can concentrate on transcribing rather than worrying too much about content, i.e. the process of creating ideas for writing is done before the child tries to actually write them. This avoids the problem of the child trying to do two things at the same time: generate the ideas for the story plot and think about how to transcribe them. That challenge can be overwhelming and lead to writer's block.

The system is therefore particularly useful when children are just beginning to write.

A key link between storytelling and writing is shared writing, where the teacher demonstrates aspects of writing down the story before the children have a go themselves. Usually, this demonstration is participative and includes the class generating ideas and thinking about which idea is best for a particular purpose.

The process described here, which covers the main elements of the writing process, can be summarised as: generate, reflect, select:

- **generate**, where the writer has a number of ideas about what to write next;
- **reflect**, where the writer thinks about which one will best suit the writer's purpose;
- **select**, where the writer chooses the one they prefer.

A typical shared-writing teaching sequence, after learning and deepening *The Billy Goats Gruff*, might look like this:

Typically the teacher will first 'box up' the text into sections, and establish a purpose for each section like in the example below so that the class understands the structure. Sections (which may be paragraphs) can be shown with words or with pictures.

Section	**Purpose**
Three goats decide to go on a journey	Establish purpose, characters and setting
Three goats stop at the bridge	Establish dilemma and dangers of the troll
Two goats cross the bridge and trick the troll	Show courage and cleverness plus the foolishness of the troll
Third goat butts the troll out of the story	Strength, victory, hooray!
Three goats eating sweet grass	Happiness, safety

The teacher may then demonstrate writing for one or more sections, creating a written text on a flipchart or whiteboard by asking the class for suggestions for words or sentences to include in the text as it develops.

In this way the teacher demonstrates the 'generate, reflect, select' process by creating a text with the class, hence 'shared writing'. More details and explanations can be found in the chapter on 'Shared Writing' in *The Storytelling School Handbook for Teachers* (Smith and Gillian, 2014, pp. 77–93).

What about non-fiction?

Non-fiction, sometimes called 'non-narrative', covers all kinds of communication that are not fictional stories. It includes things like:

persuasion | explanation
discussion | recount
instruction | information reports

These are all essential communication skills for life, not just for writing but also, crucially, for speaking. They encompass the skills needed for work, relationships and leisure. It's really important to learn these skills at school, especially if they are not modelled and practised at home.

Non-fiction pieces are similar to stories in that they can be rehearsed orally and they all have a structure with a beginning, a middle with clumps of information, and an ending. The same method that children use with fiction can be applied to these non-fiction pieces: telling (HMSS), deepening, innovation, invention and links to writing through shared writing.

Sometimes a non-fiction piece can be linked to a story, allowing rehearsal of the structure using the fiction content.

For example, with *The Billy Goats Gruff*, non-fiction pieces, both oral and written, might include:

persuasion: the older goat persuades the younger to go first;

discussion: should they cross the bridge or go home?

recount: how the goats got to the green grass;

factual reports: about goats and trolls.

Such pieces can be learned orally using HMSS, deepened, and also developed into written pieces using the same kind of sequence as with fiction.

How does it all fit together?

Here's that diagram again showing how the elements can link up:

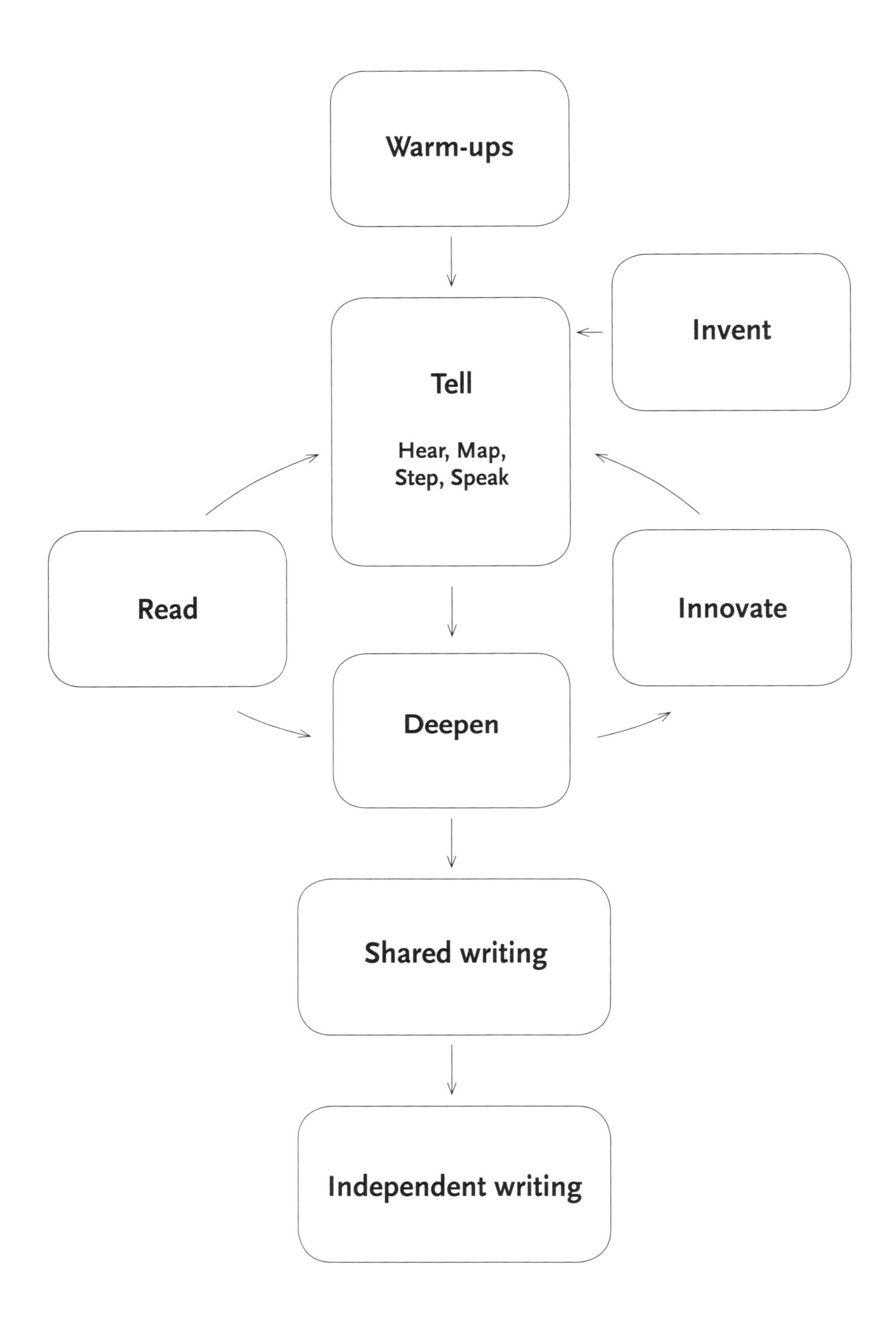

Many Storytelling Schools organise their literacy teaching by 6–8 week terms. In each of these terms there is often a focus on one fiction story and one non-fiction piece. These may be linked to a common topic, or the non-fiction piece may be linked to the content of the story in some way. In this way a story curriculum is constructed across the school.

So, if you are working in a Storytelling School you can include some of the stories in this book in your story curriculum every year.

Some schools have each term a lead topic that most of the teaching is linked to. If you are in such a school you can choose stories from this book which fit with science-related topics in a given term, and use them to teach literacy too.

In other schools you may want to use these stories simply as a springboard for science teaching. Various ways to do this are spelt out in the teaching ideas listed after each story in the book.

How to use this book

This book contains lots of ideas for ways to teach a story to your class and then link the story to the teaching of science, literacy and other areas of the primary curriculum. We imagine it being used something like this:

1. First choose the science topic that you want to teach and check that the story or stories can be adapted for the age and ability of your class.

2. Learn to tell the story yourself and then teach your class to retell the story, using usually the HMSS method.

3. Next you can choose some options for deepening the story, including discussion, role-play, re-enactment and so forth.

4. We then offer you a number of ways to explore the science in the story through explanation, discussion and practical science activities. Choose the ones most suitable for your class and your teaching aims.

5. Finally, each topic comes with some ideas for using the story as a springboard for teaching literacy and other curricular areas. Again choose the ones most suitable for your teaching needs, or make up your own!

So that's our brief summary of the Storytelling School method. There's much more detail in the rest of the handbook where we show you how to apply these ideas to science stories.

Chapter 3

Biology Stories

Topic 3.1 SEEDS AND SEASONS

Story: Anna's Apple Tree

Topic: Seeds and seasons
Plot type: Rebirth
Genre: General fiction

This is a story about a little girl and her grandfather's apple tree. As she grows, he teaches her about apples and seeds. Then, after his death, she plants a seed from his old apple tree and nurtures it into a new tree. Later, she teaches her own children about seeds and trees.

The science content in the story includes the life cycle of an apple tree, which produces fruit whose seeds become trees. This cycle is echoed in the births and deaths in the girl's own family. The story also covers sequences relating to the seasons and the care of plants.

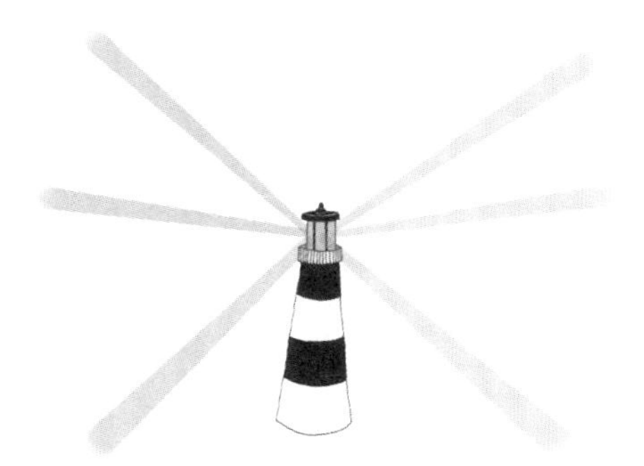

Anna's Apple Tree

Once ... there was a little girl called Anna who loved her grandfather. She visited him every week in his cottage. He would take her out into the garden and show her all the trees. Her favourite was the apple tree with its old knotted branches.

Every summer, when the tree was in full leaf, Grandfather would pick Anna up and put her into the branches of the trees. Then he'd step back and play peekaboo through the shiny green leaves. At the end of the day, he'd give her a big juicy apple for the long walk home.

One summer day, Anna went to visit her grandfather.

'Can I have an apple tree, Grandfather? Can I have one in my garden?' she asked.

'Of course, my love. Here, have some seeds!' he replied, and he handed her a rosy apple from the tree.

Anna looked at the apple in surprise. 'In here?' she asked. 'The seeds are in here?' she said as she took a big juicy mouthful.

'Yes!' laughed Grandfather. 'Plant the pip and you'll get a brand new apple tree. But remember – good things take time to grow, just like you!'

Anna was amazed.

Not long after, in his sleep, her grandfather died. Anna was so sad. She missed him so much. After the funeral she went to visit his house to see the garden and the old apple tree, but when she got there some men were cutting it down.

'Stop!' she shouted. 'That's my grandfather's tree. Why are you cutting it down?'

'Sorry love,' said a man. 'The tree is old and dying. It has to go.'

Anna picked up one of the last few apples and went home and sat on her bed, remembering all the times with her grandfather, turning the last apple over and over in her hands. Then she remembered the seeds.

'I know,' she thought. 'I'll grow a new tree from the old seeds.'

She took the apple and cut it open. Inside were five shiny brown pips.

She planted the seeds in her garden in a row and went to bed.

And then came the rainy, rainy autumn. Anna went out to see if her seeds had grown but there were no apple trees.

And then came the cold, cold winter. Anna missed her grandfather and his garden and the apple tree, so she went out to see if her seeds had grown ... but there were no apple trees.

And then came the hazy, hazy sunshine of spring. It was a little warmer. The garden began to look green. Anna missed her grandfather and his garden and the apple tree, so she went out to see if her seeds had grown but ... there were no apple trees.

She went to find her mother and she cried. 'I wanted a tree to remember my grandfather. I planted five seeds but no trees have grown.'

Her mother said, 'Remember that good things take time to grow, just like you. Show me where you planted the seeds.'

Anna showed her mother. There in the ground was one little green shoot.

'Look!' said her mother. 'Here's your apple tree. It's tiny now but, if you care for it, one day it will be a great big tree like the one in Grandfather's garden.'

'Really?' said Anna

'Yes,' said her mother.

And then came the hot, hot summer. Anna watered her tree every day and watched it grow tall in the warm summer sun. It grew a strong little trunk, as wide as her finger, and five shiny green leaves. It was too small to sit in but Anna loved it. It was a tree to remember her grandfather.

And then came the rainy, rainy autumn again and the five shiny green leaves turned brown and fell to the ground. But Anna didn't mind; she knew it would grow more next spring.

The seasons changed, winter, spring, summer, autumn ... winter, spring, summer, autumn... and so the years passed.

Anna grew taller and the tree did the same. When Anna was ten years old the tree had its first flowers. She watched as bees came and visited the flowers and then, a few weeks later, three tiny apples started growing! Anna was so excited.

'Look Mum!' she said. 'Apples from Grandfather's tree!' The apples grew bigger and bigger and in autumn she picked them. They ate them together after supper.

'Well done,' said her mum. 'Grandfather would be proud of you!'

The tree grew strong and tall. Anna did the same. The tree gave more and more apples every summer. Every time Anna picked one, she remembered her grandfather.

The years passed and Anna left home and got married. Soon she had a daughter of her own called Pip. When Pip was just three years old, Anna took her to visit the tree in her mum's garden. It was much bigger than Anna now and full of ripe apples. She sat her daughter up in the branches of the tree and they played peekaboo for a while.

Then Anna said, 'You know something, Pip? When I was little, my grandfather played this game with me in his apple tree.' She told Pip about her grandfather and Pip listened wide-eyed. Anna picked her an apple from the tree and said, 'If you like, you can grow an apple tree of your own, just like I did. But remember, good things take time to grow – just like you.'

They took the apple home and planted the seeds in their own garden.

Top tips for telling

The heart of the story is the love of Anna for her grandfather and how growing a new tree helps her honour his memory and cope with her loss. When you tell the story emphasise how much they love each other and how lost she feels first when he dies and then when their favourite tree is cut down. After that, the uncertainty of whether the seeds will grow provides some drama and then there is the satisfying completion when Anna teaches her own child.

When you tell the story you can cut open a real apple with a knife at the right time and show the class the seeds inside.

Ways to work with the story

Teach the story using HMSS

Deepen the story with activities such as these:

- Practise telling the sequence from seed to fruit by drawing a map of the sequence, stepping the sequence and telling the sequence, just as you would do to learn a story using the HMSS process (see p. 14). You could do the same for the journey from baby to parent.
- Role-play the mother's advice to Anna through the seasons as she nurtures her tree.
- Role-play Anna explaining about seeds and trees to her daughter.
- Act out the sequence of the seasons using actions to represent the changing weather in mime or dance.

Explore the science that is directly linked to the story:

- how seeds grow into plants (including planting seeds in class);
- what plants need to grow and stay healthy (light, air, water and soil);
- naming parts of plants and trees;
- how plants change over the seasons;
- how the weather changes over the seasons;
- life cycles of plants and humans.

Try these science activities:

SEEDS AND SEASONS

1. Investigate germination

Try soaking seeds to see them swell up. Beans swell up quite a lot before they germinate so that the seed coat softens and splits. You may want to see what happens when you soak other seeds. Chia and Basil seeds change from tiny black dry grains to something that looks like frogspawn!

2. Planting seeds

Plant bean seeds in the classroom. You can even grow them up strings leading to a picture of a giant's castle. Plant most in soil, but it's a good idea to plant one in a clear pot and pressed up against the pot's edge so that the developing seedling can be seen. Photograph the seedlings once a week or use a time-lapse photography app to make a film of one growing. You may want to try planting some seeds in odd places, e.g. in fabric or in a bowl of rice to show that they need water to grow.

3. Identify trees

Go on a walk to a park and learn how to recognise common trees by looking at their leaves. Talk about trunks and roots and branches and leaves.

Also note that some trees do not lose their leaves in autumn. Use pictures and simple keys to identify the trees from the shape of their leaves.

4. Label the parts

Learn the names of the parts of the trees and other plants. You could do this by making large collages of trees and plants you saw on your walks, using leaves you've collected and/or bark rubbings. Then ask the children to write labels for the pictures.

5. The seasons

Play the seasons – bring a range of clothes and accessories that symbolise the weather in the different seasons. Create a photo collage of pictures of the children dressed for spring, summer, autumn and winter and each doing something relevant to the season, as suggested by the child, e.g. acting out Bonfire Night in the autumn, wearing coat, hat and boots because it's wet and windy and gets dark early.

6. Life stories

Ask the children to bring in a selection of photographs of themselves at different stages of life. Look at pictures of an adult (it could be you) throughout their life and show how these stages form a life cycle. It's also fun to invite a family into the class if you can find one with three generations – a granddad, mum and baby. Get the students to explain how the family has grown and how the children will grow up and the parents become grandparents.

7. Life cycles

Draw the developing plants at various stages and show how these form a life cycle. Look at other life cycles that are different, e.g. chicken, frog or butterfly.

Explore other curriculum areas that link to the story:

Reading

Read other stories including plant growth, e.g. *Jack and the Beanstalk* told by Kathleen Lines in *Jack and the Beanstalk: A Book of Nursery Stories* (Oxford University Press, Oxford, 2013), *The Gigantic Turnip* by Aleskei Tolstoy (Barefoot Books, Bath, 2006) or the folktale *Apple Tree Man* (see: thecompanyofthegreenman.wordpress.com/2009/01/23/the-apple-tree-man).

Fiction

Box up and write the story, or innovate using other trees or plants.

Poetry

Write seasonal poems with four verses and a given season for each verse. You could base them on the changes seen on a deciduous tree – losing its leaves for the winter and growing a new coat of leaves in the spring.

Non-fiction

Instructions: Write or sequence instructions on how to grow/care for trees.

Instructions: Write a recipe and instructions for making an apple pie.

Instructions: Write or give oral instructions on how to grow a magic beanstalk.

Explanation: Explain the life cycle of a tree to a friend. Once the children have learned how to explain the life cycle of the apple tree, it is easy to use the same basic structure to innovate an explanation of the life cycle of another fruit tree.

Information report: Write a report on trees in our school.

Information report: Write a report entitled 'Where Are the Seeds in Different Fruits?'

Art

Paint pictures of the changing seasons.

Use leaves to make prints or collages.

Slice open fruit and sketch what you can see inside.

Study the work of Andy Goldsworthy, who uses natural objects to create sculptures that are transitory, changing as they decay or melt away.

Religious Education

Investigate harvest celebrations from around the world.

Topic 3.2 SENSES

Story: Going to Catch a Piggy-wig

Topic: Senses
Plot type: Rebirth
Genre: General fiction

This story rhyme tells how the wolf uses all his senses to sneak up on a piggy-wig to catch him for his supper. There are two possible endings. In one, the piggy-wig gets away. In the other the wolf gets to taste the piggy-wig and crunch up his bones.

The science content in the story is an introduction to the names of all the senses and how we use them.

Going to Catch a Piggy-wig

Imagine you were a big hungry wolf...

Going to catch a piggy-wig
And eat him for my lunch
Hold him by his curly tail
Crunch crunch crunch

Going to catch a piggy-wig
I can hear his snuffles
Snorting with his piggy snout
Digging up the truffles

(Snuffle snuffle snuffle snuffle
Digging up the truffles)

Going to catch a piggy-wig
I can smell the pong
Rolling in the stinky muck
All day long

(Stinky stinky stinky muck

All day long)

Going to catch a piggy-wig

I see where he goes

Let's creep up behind him

On our tippy toes

(Creep creep creep creep

On our tippy toes)

Going to catch a piggy-wig

I feel his hairy skin

Scratchy like the stubble

On a daddy's chin

(Scritch scratch scritch scratch

On a daddy's chin)

Going to catch a piggy-wig

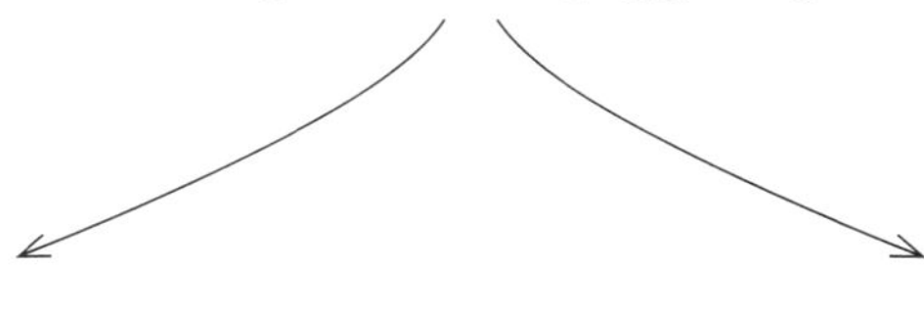

ENDING 1

Going to catch a piggy-wig

I can taste him with my tongue

AAAH AAAH TISHOOOO

Where's my piggy gone?

ENDING 2

I can taste his meat

His nose tastes quite delicious

And now I'll try his feet.

Going to catch a piggy-wig

I crunched up every bone

My belly's full of piggy-wig

So now I'm going home!

Top tips for telling

This poem is to be learned word for word and told in the voice of the wolf, so find a deep scary voice for the wolf to build up the drama. You can learn the story by mapping and stepping each verse, and then telling it with lots of actions for the children to join in with, and maybe chanting the whole thing together as a class. The mapping and the actions you devise while stepping the poem will also help you remember what happens next. Emphasise the sense organs involved. Make your actions big and dramatic.

Ways to work with the story

Teach the story using HMSS

Deepen the story with activities such as these:

- Explore in mime or dance the way we use our senses. You could be a flower fairy checking that all the roses have the right scent, or Mr Wonka in the chocolate factory, checking the taste of all the sweets (as in Roald Dahl's *Charlie and the Chocolate Factory* [Puffin Books, London, 2007]).
- Re-enact the poem with plenty of actions.
- Role-play the piggy-wig telling Mummy all about how he was nearly eaten by the wolf. Tell her about each sense that the wolf used to find him.
- Role-play being the wolf telling his mummy about stalking the pig.
- Make up a senses poem about going to buy an ice cream (or another food that has associated sounds): hear the van, smell the flavour, see the colours, taste the sauce, feel the cold.
- Write a list poem using the five senses: I can see ... I can hear ... I can feel ... I can taste ... I can smell.

Explore the science that is directly linked to the story:

- the senses.

Explore the science that is indirectly linked to the story:

- sorting and naming animals;
- what an animal needs to survive;
- parts of the body.

Try these science activities:

SENSES

1. Name the body parts

Sort animals into groups according to their features – shells, wings, legs, fins, tentacles, etc. – noting that lots of animals have, like us, four limbs, two eyes, a nose, a mouth, etc. Play body part naming games and sing body songs.

2. Sight

Look at photographs of eyes on different creatures. How are they different? Look at where the eyes are on the head – hunters need both eyes at the front to have better forward vision. Prey need to see what is behind them so their eyes are on the sides of their head. Play games in which the children have to draw or move safely while blindfolded. Help the children to realise that we rely on our other senses in the absence of sight.

3. Hearing

Look at photographs of ears on different animals. Find out if all animals have ears. Go on a sound walk and collect sounds from around the school on a video recorder. Can the rest of the class identify the sounds by listening to the video, without seeing it? Prepare shakers with different things inside for the children to identify by sound alone.

4. Touch

Ask the children to identify different surfaces with the back of their hands and then again with their fingertips. Discover which side of their hands is most sensitive. Find out what babies use to feel (lips) and what animals use to feel (noses, whiskers, antennae, etc.). Use fingertips to read braille – can you recognise a word? Make a feely picture in which the textures vary so the shapes can be felt. Make feely bags: conceal objects inside sealed, opaque bags and have the children try to identify what is inside by feel alone.

5. Taste and smell

Have a smelling and tasting session. We like to use chopped vegetables and fruits of various sorts and bite-sized pieces of different breads. You can ask the children to be professional tasters for a shop and decide which is the best fruit/veg/bread to sell, or link the activity to making sandwiches/a picnic in Design/Technology.

6. Looking after an animal

Introduce an animal into the classroom for a week. Make notices to put around the cage with instructions about how to care for the animal, or record the instructions in the children's books by asking them to annotate a photograph of the animal in its cage. For example, 'Provide plenty of water to drink' could be an annotation for the water bottle. Make sure you cover breathing, eating and drinking. A guinea piggy might be fun!

Explore other curriculum areas that link to the story:

Reading

Read other versions of *The Three Little Pigs*, e.g. *The Three Little Wolves* and the *Big Bad Pig* by Eugene Trivizas (Egmont, London, 2003).

Read other wolf stories, e.g. *Little Red Riding Hood*.

Read other traditional tales told in rhyme e.g. *Little Red Riding Hood and The Wolf* by Roald Dahl (A & C Black, London, 2005).

Fiction

Innovate using different animals and using all the senses – it doesn't need to rhyme.

Non-fiction

Instructions: Write instructions on how to avoid being eaten by a wolf. This could be imaginative rather than factual, e.g. dress up as a woodcutter, or cover yourself in toothpaste because wolves hate cleaning their teeth.

Instructions: Be a hunter giving his son instructions on how to catch a wolf.

Instructions: Write a page for a recipe book: how to make a piggy-wig sandwich.

Information report: Write a fact file entitled 'Wolf Facts'.

Recount : Write a letter home to Mummy Pig telling her how you narrowly escaped being eaten by a wolf.

Persuasion: Persuade Mr Wolf to eat a different animal/food instead.

Persuasion: Create a 'wanted' poster for Mr B.B. Wolf.

Design/Technology

Prepare a healthy lunch. This could be making sandwiches or you could make different healthy foods in groups for a class picnic.

Design/Technology and Music

Make musical instruments that all sound different. Make one that is loud/quiet or makes a high/low sound. Make one that you pluck like a guitar or beat like a drum. Use the instruments to tell the story.

Life Skills

Think about how we stay safe. Discuss rules. Red Riding Hood was told to stay on the path for a reason! Think about how rules keep us safe.

Invite an optician/doctor/nurse into class to talk about looking after our eyes and ears.

Topic 3.3 HUMANS, HEALTHY DIETS, TEETH AND BONES

Story: Uncle Jack

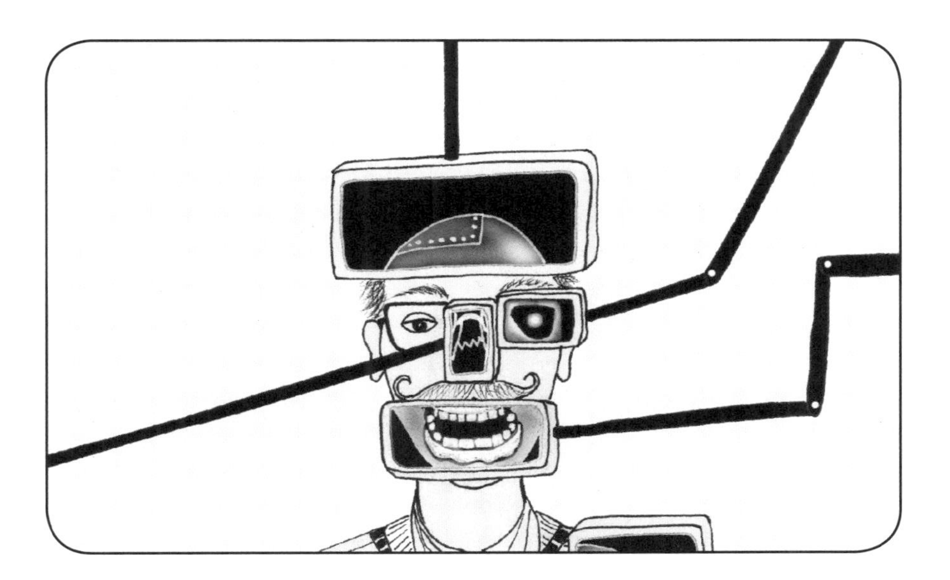

Topic: Humans, teeth, bones and healthy diet

Plot type: Comedy

Genre: Poetry

This is a poem about Uncle Jack who has not taken care of his body during his lifetime. The poem maps the sequence of his accidents and illnesses in chronological order. The message of the poem is a warning to the young to take care of their bodies because they won't get another one.

The science content in the story includes caring for our teeth, the names of some key bones and organs, and the damage that some foods can do to our bodies. If you do the poem with actions, you can help the children to locate the body parts mentioned in their own bodies.

Uncle Jack

My Uncle Jack, with the twirly moustache, said,

'There's practically nothing left of the body I was born with. You're all new and shiny. Take care of what you've got. You won't get another one.'

When he was one he punctured a lung.

When he was three he broke his knee.

When he was four, they started a war and Hitler blew up the house next door.

Uncle Jack was knocked out cold, knocked out cold at four years old! When he awoke ... his nose was broke.

When he was eight, he fell off a gate, left his skull in an awful state ... They had to fix it with a metal plate!

When he was older he dislocated a shoulder.

He drank too many coffees and ate sticky toffees.

'They pulled out every tooth in my head' he said.

He had marble eyeball, a dicky ticker*, and a box to make his heart beat quicker, a peg to mend a broken leg when he slipped downstairs in his favourite slippers.

At sixty-six they replaced both hips.

'They say man is made of mostly water but I can't be more than a quarter!

I must be mostly made of plastic. Isn't that fantastic?' said Uncle Jack.

* *A 'dicky ticker' is a heart that is weak and not very healthy.*

P
PHOTOCOPIABLE FOR USE IN SCHOOL

Top tips for telling

This is another poem that you might get the class to learn word for word. It can also be told in a freer and improvised way as a soliloquy in which Jack recounts his story of body part replacement.

To learn it word for word, you can tell it with a lot of actions (gestures or mime that describe the ideas in the poem). This will help both you and the class to remember the lines. You can tell it with call and response for increasingly longer sections of the poem with the class standing in a circle and echoing back the words and movements.

As this is a big sequence, you could map the sequence of body parts to make it easier to remember. You could add some key words on each part of the map, since this will help too, as will stepping each line.

Apart from that, the main thing is to make the whole thing as much fun as possible.

Topic 3.3 HUMANS, HEALTHY DIETS, TEETH AND BONES

Alternative story: The Smiling Princes

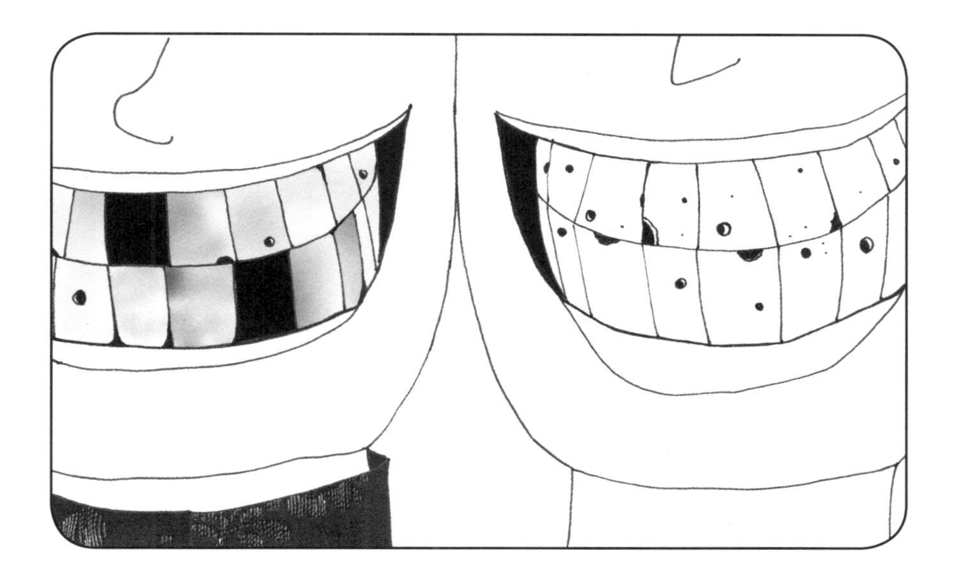

Topic: Healthy teeth
Plot type: Rags to riches
Genre: Folktale

This is a story about a fussy princess who wants a husband with a bright smile and sweet-smelling breath. Three princes propose. The first never brushes his teeth, the second only eats toffees but the third brushes his teeth and eats well. She marries the prince with the good teeth!

The science content in the story is about how diet and brushing affects our teeth.

The Smiling Princes

Once there was a princess who lived with her father in a grand palace. One day the King said, 'It's time for you to marry. What kind of husband would you like?'

She said, 'Father, I would like a husband with a nice smile, that's all. I'd like a man with a smile as white as snow, breath as fresh as mountain air and teeth that will last as long as he lives!'

In the palace over the mountain, there lived three princes. The eldest had never brushed his teeth. Not even once!

He said, 'I think I will go and marry the princess.'

So, the eldest prince climbed over the mountain to the grand palace and asked to see the princess. He fell on one knee and with a great big smile he said, 'Will you marry me?'

The princess was shocked. The prince's teeth were in a dreadful state.

'YUCK. CERTAINLY NOT!' she said. 'Your teeth are brown and your breath smells like rotting fish! Go away!'

Sadly the smelly prince slunk off back home over the mountain.

'She said no!' he said to his brothers. 'My teeth are too brown and smelly!'

'Let me try,' said his younger brother. 'My teeth are nice and clean. I brush them every day!'

The second prince did indeed brush his teeth twice a day, but he was always scoffing toffee, chocolate and fudge. He never went anywhere without some in his pocket.

He climbed over the mountain to the grand palace and asked to see the princess. He fell on one knee and said, through a mouthful of fudge, 'Will you marry me?'

'Open your mouth!' she said and looked inside.

'YUCK, CERTAINLY NOT!' she said. 'Your teeth are nice and white and your breath smells alright, but your teeth are full of holes. They won't last long. I will not marry a man whose teeth will fall out! Go away!'

Sadly the prince, chewing toffees all the way, trudged home over the mountain.

'She said no!' he said to his brothers. 'My teeth are too full of holes!'

'Let me try!' said the youngest prince. 'Maybe she will marry me.'

The youngest prince brushed his teeth twice a day, ate cheese and nuts and fish and never, ever ate toffee, chocolate or fudge.

He climbed over the mountain to the grand palace and asked to see the princess. He fell on one knee and said, 'Will you marry me?'

She looked at his smile, and it sparkled white as snow.

She sniffed at his breath and it was fresh and clean as mountain air.

She looked at his teeth and there was not a single hole or filling.

'OOH YES!' she squealed. 'I will marry you! Your teeth are nice and white, and your breath smells fine, and your teeth are strong and good. You are the man for me!'

And so they were married.

Every morning they sang this song (to the tune of 'Row Row Your Boat'):

Brush your teeth, up and down
Morning and at night
That's the way to brush your teeth
To keep them clean and white

Brush your teeth, up and down
Do it twice a day
Then they'll still be in your mouth
When you're old and grey

They brushed twice a day, they ate good food and never toffee, and so their teeth stayed strong and free from holes until the day they died.

Top tips for telling

This is a funny story so ham it up. You can make the princess fussy and funny as she wishes for her perfect husband and rejects the first two suitors. Make the first two princes enthusiastic but doomed as they propose. Give the third prince a nice bright smile and cool confidence.

The song reinforces the message so don't forget to sing it!

Ways to work with these stories

Teach one of the stories using HMSS

Deepen the story with activities such as these:

- Explore the parts of the body by learning the Uncle Jack poem and moving with it to make a dance.
- Re-enact the poem, bandaging Uncle Jack in all the right places as each accident happens.
- Role-play a doctor/dentist telling Jack why he is hurt, what is broken and how he can avoid it happening again.
- Role-play Jack phoning his wife from hospital to explain why he is there again.
- Role-play the princess explaining why she won't marry the princes.
- Role-play the royal dentist telling the princes how to look after their teeth.

Explore the science that is directly linked to the stories:

- eating the right types of food keeps our body and teeth healthy;
- our skeleton and muscles give us support, protection and movements.

Explore the science that is indirectly linked to the story:

- the different types of teeth and their functions;
- what happens when we eat.

Try these science activities:

HUMANS, HEALTHY DIETS, TEETH AND BONES

1. Body brainstorm

Brainstorm what the children already know about the human body. Children often have lots of little bits of knowledge about the body. A good way to tie all this together is to give out large sheets of paper with a basic human outline drawn on it and ask them to draw what's inside that we can't see. They will often know about bones, blood vessels and the heart and brain. Use these drawings to introduce what is actually inside us. Name some of the major parts of the human body and play games like 'Simon Says ... touch your skull!'

2. The skeleton

Investigate the places where we can bend and twist by trying some yoga moves and observing whether our joints open and close like hinges or rotate fully. Identify and name the bones in the human body. Compare animals with exoskeletons with those which have endoskeletons to see how they both give support and protection. Mark all the bones that Uncle Jack broke on a picture of a skeleton.

3. Healthy diets

Read the labels on food products to find out what is in them. Look at the proportions of different kinds of foods that we need to stay healthy. Design and make a healthy packed lunch.

4. What happens when we eat?

It can be fun to start off with what we can learn from our digestion going wrong! Children are familiar with vomit, constipation and diarrhoea, so use them as topics to begin a discussion. Once the children have got over the inevitable surprise, they often have a lot of questions. The acidity of vomit shows us how acidic our stomach is and that we need this acid to break down our food so that nutrients can be absorbed from the upper intestine. Constipation and diarrhoea show us that our lower intestine must take out just the right amount of water for us to be healthy. Then learn to sequence the main structures in the digestive tract: oesophagus, stomach, upper intestine and lower intestine. You could even map and step this process.

5. Our teeth

Find out about names of teeth and their functions. Give the children a mirror so they can count their own teeth. Investigate drinks that are bad for our teeth – you can put eggshells into different drinks for a week and see which dissolve the shell most quickly. Invite a nurse into school to talk about dental care.

6. Changes

Look at how we have changed and grown since we were born. Discuss all the things we do for ourselves to keep us healthy, e.g. washing our hands and brushing our teeth. If you can find a willing participant, invite a mother and baby into the class to see how much the mother has to do to keep the baby healthy: sterilising bottles, changing nappies, using safety seats and special bedding, etc.

Explore other curriculum areas that link to the stories:

Reading

Read other stories and poems about looking after our bodies, e.g. *Oh I Wish I'd looked After Me Teeth* by Pam Ayres (in *The Works: The Classic Collection*, Ebury, London, 2008).

Listen to Chris Smith telling the story of *'The Skeleton Woman* (www.storymuseum.org.uk/1001stories/detail/102/skeleton-woman.html).

Fiction

Box up and write the story, or innovate using different ailments for Uncle Jack. It doesn't have to rhyme.

Innovate a new story about three princesses who try to win the hand of a prince.

Innovate a new story about three footballers who want to be picked for their team but only one can show that he looks after his/her health.

Poetry

Learn and perform some poems about body parts, e.g. *Boa Constrictor* by Shel Silverstein (in *Falling Up*, HarperCollins, London, 2006).

Non-fiction

Instructions: Write a pamphlet for Uncle Jack, from the dentist, about how to take care of his teeth ... or dentures!

Information report: Write a report on a science experiment – 'Which Drink Is Worst for Our Teeth?'

Information report: 'Skeleton Facts'.

Recount: Write Jack's diary for the day when he fell down the stairs.

Explanation : Write a pamphlet on why we should eat healthy foods.

Persuasion: Be the dentist and persuade Jack to stop eating toffees.

Persuasion: Create an advert for a new toothpaste.

Art

Draw family portraits, sketching from photos or real life.

History

'Uncle Jack' links easily to the topic of World War II and the effect of food rationing on health.

Design/Technology

Taste healthy foods and prepare healthy meals. Invite an optician/doctor/nurse into class to talk about looking after our eyes and ears.

Topic 3.4 HABITATS

Story: Mummy, Can I Have a Penguin?

Topic: Animals and their adaptations to their habitats
Plot type: Voyage and return
Genre: General fiction

This is a story about a little girl who wants a penguin. As her house is not the right habitat for a penguin, she journeys to Antarctica to be with the penguins there. She soon finds out that Antarctica is not the right habitat for her and she comes home.

The science content in the story concerns the different habitats needed for a rabbit, a duck, bees and a penguin. It also considers how humans find it hard to live in extreme habitats.

Mummy, Can I Have a Penguin?

Laura wanted a penguin more than anything in the world. She asked for a penguin every birthday and she asked for a penguin every Christmas.

'Mummy can I have a penguin? I want a penguin called Snowy to come and live in my bedroom.'

And every year she got the same answer. 'You can't have a penguin because your bedroom is just not the right place. If you want a penguin as a pet you'll have to move to Antarctica and live with the penguins on an iceberg and eat fish for your supper.'

But Laura still wanted a penguin more than anything in the world.

Instead of a penguin, her Mum bought her a rabbit.

'Can I keep it in my bedroom?' Laura asked.

'No, Laura, rabbits live outside in the garden where they can dig in the soil and eat the grass. You can't keep a rabbit in your bedroom. It's just not the right place.'

Laura liked the rabbit. She called it Snowy, even though it wasn't white. And she looked after it in the garden pretty well.

But she still wanted a penguin.

'Mummy, can I have a penguin?' she asked.

She still got the same answer. 'You can't have a penguin because your bedroom is just not the right place. If you want a penguin as a pet you'll have to move to Antarctica and live with the penguins on an iceberg and eat fish for your supper.'

Instead of a penguin, her Mum got her a duck.

'Can I keep it in my bedroom?' Laura asked.

'No, Laura, ducks live outside in the garden where they swim in the pond and eat weeds. You can't keep a duck in your bedroom. It's just not the right place.'

Laura liked the duck. She called it Snowy, even though it wasn't white. And she looked after it in the garden pond pretty well. The duck was very happy swimming around in the pond, eating the weed. It even made a nest.

But she still wanted a penguin.

'Mummy, can I have a penguin?'

She still got the same answer. 'You can't have a penguin because your bedroom is just not the right place. If you want a penguin as a pet you'll have to move to Antarctica and live with the penguins on an iceberg and eat fish for your supper.'

Instead of a penguin, her Mum got her a beehive.

'Can I keep it in my bedroom?' Laura asked.

'No, Laura, bees live outside in the open air. They like to buzz around and find flowers and make honey. You can't keep a beehive in your bedroom. It's just not the right place.'

Laura liked the bees. She called them all Snowy, even though they weren't white. And she looked after them in the garden pretty well. The bees were very happy buzzing around in the garden, visiting the flowers. They even made some honey.

She still wanted a penguin.

'Mummy, can I have a penguin?'

She still got the same answer. 'You can't have a penguin because it's just not the right place. If you want a penguin as a pet you'll have to move to Antarctica and live with the penguins on an iceberg and eat fish for your supper.'

So, one day, Laura packed her warmest jumper and her thickest socks and went to Antarctica. And she found some penguins. She even found one who seemed to like her and would take a fish from her hand. She loved her little penguin pet. She called it Snowy! She fed it fish and went ice sliding and had lots of fun. She built herself an igloo and she lived there on the iceberg, in the cold, among the penguins. Every day, she ate raw fish for supper.

But at night, even in her thickest socks, her toes were cold and she missed her warm bed. And she soon became fed up with raw fish for supper every day.

One day she said, 'I don't think I can stay here any more. Its just not the right place ... for me.'

So she said goodbye to Snowy and packed up her socks and jumpers and went home.

That night, after she had eaten baked beans on toast and her mummy was tucking her toes into her nice warm bed, Laura said, 'Mummy, I want to have a penguin but I can't have a penguin because it's just not the right place. And I don't want to move to Antarctica and live with the penguins on an iceberg and eat raw fish for my supper. Lauras live in houses with their mummies and have nice warm beds and eat baked beans on toast. This is the right place for me.'

Top tips for telling

The character of Laura is central to the story. She REALLY wants a penguin and goes on and on about it! Maybe make her a bit annoying and very determined.

Set the scene with the rabbit, the duck and the bees. This can have a comedy edge, with the class joining in or taking over the telling, maybe thinking up other animals to add. Also, don't forget to hop, flap and buzz when you tell this part.

Then comes the big moment when Laura goes off to the Antarctic. Suddenly, we are in a fantasy adventure story. Evoke the wonder of living with the penguins for a while, then the love and comfort of her homecoming.

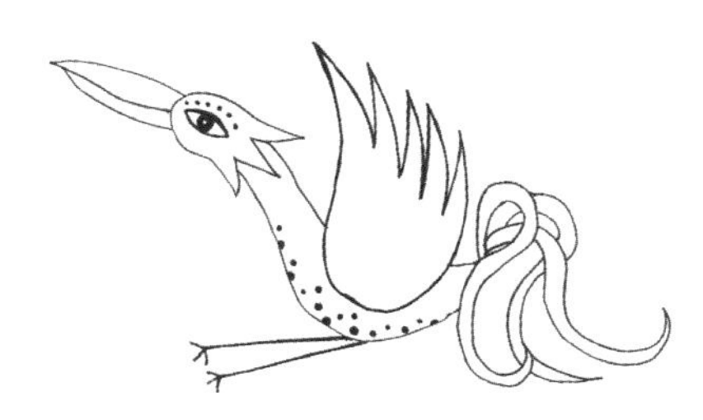

Ways to work with the story

Teach the story using HMSS

Deepen the story with activities such as these:

- Re-enact the story, being all the different animals in their different habitats.
- Role-play going to Antarctica. Pack your bag full of things to keep you warm and to help you survive in this very different environment.
- Role-play being Laura's mother and explaining why Laura can't keep a rabbit (or squirrel or shark) in her bedroom.
- Role-play being a penguin expert/keeper. Explain how a penguin is suited to life on the ice and what kind of habitat a penguin needs.

Explore the science that is directly linked to the story:

- most things live in habitats to which they are suited;
- simple food chains.

Explore the science that is indirectly linked to the story:

- the differences between things that are living, things that are dead and things that have never been alive;
- animals and plants may depend on one another for their survival;
- the names of a variety of plants and animals in their habitats, especially local species.

Try these science activities:

HABITATS

1. Sorting things

Sorting objects/animals into groups of things that are alive, things that have been living (such as wood) and things that have never been alive. This could be done by sorting photographs of things found in the grounds or in a nearby park.

2. Meet rabbits and penguins

Invite a local pet store/farm park to bring animals to the classroom or go out and see animals in a farm park or zoo. Try to include some hands-on time with rabbits! Find out as much as you can about how to care for these animals when they are not living in their natural habitat.

3. Name living things

Make a set of cards with pictures and names of different local plants and animals. These could be cards for playing 'Pairs', in which each child takes a turn to turn a pair of cards over. If they find matching picture and name cards, they can keep the pair. The child with most pairs wins. Alternatively, make dominoes with pictures of plants (or animals) and their names instead of the dots. You could even make your own bingo game cards by laminating a set of leaves stuck onto cards and then play leaf bingo, crossing off the leaf that is called out if you have it on your card. All of these will help the children to recognise and name the species from their locality.

4. Explore habitats

Explore a habitat and identify the plants and animals you find. Aim to explore a few very different habitats, such as a pond and a log pile. Photograph or draw the living things you find and label them. Look at the structures that allow them to live in their habitat such as gills on a fish.

5. Food chains

Find out what local animals eat. If they are carnivores, find out what their prey eats. Draw simple food chains showing that all animals rely on plants. You could start with rabbits to link to the story and then look at foxes, hedgehogs, frogs and birds of prey. If you are unable to visit an outside area then using a section of a wildlife documentary may be a good starting point. If you have a pond, some simple food chains can be identified there, e.g. algae–tadpole–diving-beetle-larva. You could relate this back to the story by researching the food chain for a penguin.

6. Animal homes

Design a home for some kind of minibeast so they can be kept in the classroom for a day or two. Woodlice are hardy and will be comfortable in a plastic tank filled with dead wood and damp moss.

Explore other curriculum areas that link to the story:

Reading

Read other stories that describe animals in their habitats, such as *Little Elephant Thunderfoot* by Sally Grindley (Orchard Books, London, 2009).

Fiction

Box up and write the story, or innovate using a different animal in the garden or different animals to want as a pet (polar bear, killer whale, tropical snake).

Non-fiction

Information report: Write a report about penguins, ducks, rabbits or bees.

Explanation: Be Mum explaining why the duck should live in the garden.

Discussion: Discuss whether it is right to bring penguins to Britain.

Discussion: Discuss whether penguins should be kept in zoos.

Recount: Be Laura and send a postcard or letter home from Antarctica about meeting the penguins.

Explanation: Be Laura and explain to your mother why the Antarctic is good for the penguins but not for yourself.

Instructions: Write instructions on how to build an igloo.

Geography

Use maps of the school grounds to show where different plants and animals can be found.

Use maps of the world to show where penguins and other creatures can be found.

Find out about people who live in extreme environments, such as the Inuits.

Art

Explore the shapes and patternation of ducks, rabbits, bees and penguins.

You could look at iridescent and highly patterned insects using photographs.

Explore symmetry in animal patterns. Design symmetrical patterns for a fantasy penguin, bee or duck.

Design/Technology

Design and build a home for a small animal. You could make bird houses or hedgehog homes to put in the school grounds.

Topic 3.5 CARING FOR THE ENVIRONMENT

Story: The Drop of Honey

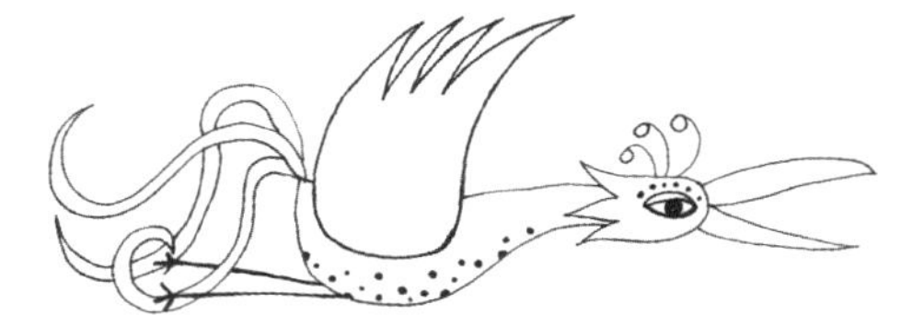

Topic: Looking after our environment, ecosystems, food chains, and extinction
Plot type: Tragedy
Genre: Folktale

This is a story about a king who sees that a drop of honey on the floor is causing a problem but chooses to do nothing about it. The honey attracts a fly, which attracts a shopkeeper's cat. The cat is killed by a shepherd's dog and the shopkeeper kills the dog. The situation escalates into war. Only then does the King see that he should have taken action when he saw the honey drip on the floor.

The moral of the story is all about taking responsibility for the world around us, and how a small problem can escalate into a large one if nobody intervenes. This moral can then be applied to the interconnections between plants and animals in a food chain or ecosystem – one tiny change can have knock-on effects and change the whole ecosystem. It can also be applied to the effect of humans on the world's ecosystem and how our inconsiderate actions can have a big effect – leaving the lights on and wasting electricity increases the amount of global warming and the extinction of endangered species.

The Drop of Honey

Once ... there was a shepherd. Every spring, he took his goats and beloved sheepdog across the river and up into the mountains of the Eastern Land. He stayed there until autumn, when the snows would drive him back west to his village. In this way the years passed.

One autumn the old shepherd, with his flock and dog, was making his way down a mountain pass in the Eastern Land, when he stopped at a village store for a rest. He had little money but asked the shopkeeper if he might buy a single spoonful of honey, as he had tasted nothing sweet for months. The shopkeeper walked over to a barrel in the corner of the shop and dipped a small metal spoon into the barrel.

Now, the King of the Eastern Land and his Chief Minister were in the habit of dressing up in disguise and wandering around the country. On that day the King and Minister were sitting drinking coffee in the same shop where the shepherd was waiting for his honey.

The King watched as the innkeeper dipped the spoon into the honey. He watched as the man walked carefully back across the room, the spoon brimming with golden amber.

The Minister noticed that, as the shopkeeper walked, the honey on the underside of the spoon was collecting together into a single drop, which was growing larger by the moment.

He whispered playfully to the King, 'Look! In a moment some honey is going to drip onto the floor. Should we tell him?'

'Not our problem!' laughed the King.

They watched as a hanging honey droplet fell down onto the stone slab next to the shepherd. The shepherd gratefully accepted the spoon, placed it in his mouth, and with eyes closed he let the sweetness dissolve onto his taste buds and savoured every sweet moment. While this was going on, a large bluebottle flew past the shepherd's legs and, catching the scent of the drop of honey on the stone floor, immediately changed course and descended towards it.

The keen-eyed Minister noticed this. 'Look, your majesty,' he whispered, 'the bluebottle is going for the honey on the floor. Should we do anything?'

'Definitely not our problem!' said the King.

The shopkeeper had a cat who hated bluebottles. She was sitting by the shepherd when she saw the bluebottle veering in her direction. Her muscles tensed and she was ready to pounce. The Minster was watching all this, and pointed it out to the King, who just shrugged and watched.

The cat jumped, claws stretched, teeth bared, soaring through the air and caught the bluebottle between her teeth, killing it instantly. But the momentum of the cat continued to carry her horizontally through the air in the direction of the shepherd's dog.

The dog saw the cat flying towards him and, believing himself to be under attack, he readied himself for a fight. When the cat landed just in front of the dog he jumped on her, sinking his teeth into her neck, killing her with one bite.

'There'll be trouble now!' whispered the Minister. 'Shouldn't we do something?'

'No need,' replied the King. 'It's not our problem. Let's just watch and see what happens.'

The shopkeeper was furious. That cat had kept his shop free from mice and rats for years. Seeing the cat lifeless in the dog's mouth, he cursed and kicked out at the dog with all his strength, connecting his boot with the dog's head. There was a crunching of bone as the dog's neck snapped.

The Minister looked at the King, but he just shrugged back. 'Not our problem!'

Now the shepherd loved that dog like his only child. When he saw its neck snap, he pushed the shopkeeper hard in the chest with his strong arms. The shopkeeper fell backwards, tripped over a box behind him, cracked his skull hard against the

stone wall and fell lifeless to the ground. Fearing for what would happen next, the shepherd rushed out of the shop and away towards his goats.

The shopkeeper's son was standing in the doorway as his father slumped to the ground. As the shepherd rushed by, he imagined that his father was dead and called to his friends outside.

'Boys! Catch him! That man just killed my dad!'

Inside the store, the Minster was getting agitated. 'Your majesty, we must do something. The youths are bringing sticks and knives. We should stop this now!'

But the King shook his head. 'Too dangerous,' he said. 'Let's wait till they've calmed down.'

Outside, a crowd of youths pushed the shepherd to the ground, laying into him with their clubs and boots until he was dead.

It wasn't long before word reached the shepherd's village that he had been killed by a mob of Easterners. Angry, the young men of his village gathered whatever weapons they could muster, crossed over the river and marched up the pass to the nearest Eastern village. They eagerly began smashing and burning whatever they could find. Soon the whole village was in flames.

Eastern soldiers were sent to restore order, but when they arrived, one of the youths pulled out a pistol and started firing. The soldiers fired back and men were lost on both sides.

In this way, a war began which neither king was able to win, yet neither felt able to stop. The war raged for ten years until, with both peoples weary of grieving for their lost sons, a truce was finally declared.

When the two kings met to discuss their truce, the Eastern King told the story of what he had seen at the shop.

'If only I had done something about the honey, or the cat, or the dog, or the shepherd,' he said, 'then maybe things would have been different.'

The kings declared that the truce day would be named Honey Day and ordered each country to remember this story as a reminder that peace is a precious and fragile thing that we should all, in our own way, protect.

Top tips for telling

In this tale the setting matters, so be clear about the two lands, the border, the river and so on. Then you need to establish the characters: the solitude and quiet, hard-working qualities of the shepherd and his love for his dog, then the observant and thoughtful Minister and the arrogant couldn't-care-less quality of the King.

The rest of the story needs precision in narration so the listeners understand how one thing led to another, and also a progressive escalation in the level of anger and fear as things get out of hand.

This is a story an audience will remember forever, with a powerful moral for us all. It's well worth a bit of preparation.

Topic 3.5 CARING FOR THE ENVIRONMENT

Alternative story: The Bird and the Forest Fire

Topic: Looking after our environment, ecosystems, food chains, and extinction
Plot type: Rebirth
Genre: Myth

This is a story about a little bird who, when faced with a forest fire, tries to save her forest home. It seems that she cannot achieve this on her own and yet she keeps trying. The Eagle God is moved by her heroic effort and his tears put out the flames, saving her home. The story illustrates how one determined action can inspire others and have a big effect.

The science content we focus on here is in the call to action. It can be applied to the fact that we should take responsibility for looking after the world around us.

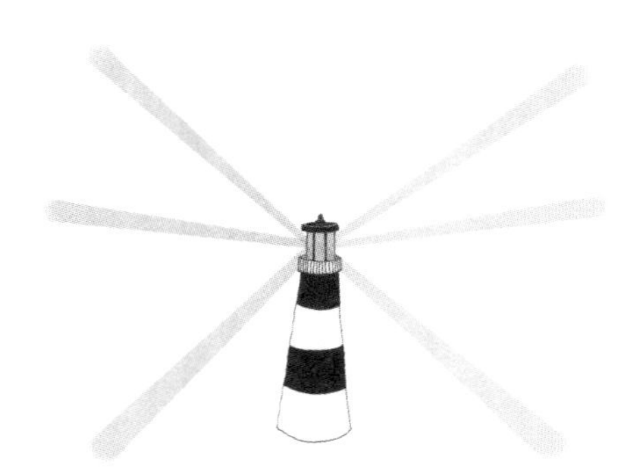

The Bird and the Forest Fire

Once ... there was a forest and in that forest there was a tree and on the tree there lived a little bird. The little bird loved her home. She loved the leaves, the branches, the tree and the whole forest. It was her home

One day, the little bird smelt something strange. She flew up, high over the canopy and saw in the distance a fire, a great snake of flames and smoke coming towards her. She called down to her friends, 'It's a fire, we must do something!' but all her friends just ran away.

'Run away with us,' they called, 'otherwise you will die!'

The little bird loved her home and would not leave it to be destroyed. She flew down to a stream by the tree in which she lived and dipped her wings into the water.

She then flew up above the fire and tipped her wings to release the droplets of water. The water hissed and fizzed away in the flames in an instant, then she flew back to the stream and again wetted her wings before returning to the fire.

She did this time and time again as the fire got closer and closer to her tree.

Up in the heavens, the gods looked down and laughed at the little bird.

'Who does she think she is?' said one.

'She'll soon be cooked!' laughed another.

But one god, the Eagle God, admired the little bird. He opened his wings, swooped down from heaven and joined the little bird as she flew between stream and fire.

'Listen, little bird,' said the Eagle God. 'You must fly away. You can't put the fire out. It's too big and you are too small. You will die.'

'I don't care if I die,' cried the bird. 'I love my home and I will do what I can to save it.'

Tears filled the Eagle God's eyes as he listened to the brave little bird. They dripped from his eyes and down into the fire, hissing and fizzing in the fire. The tears got stronger and faster until a stream of tears was flowing onto the fire. The stream became a river and soon the fire was gone, put out by the Eagle God's tears.

The Eagle God returned to heaven and the little bird to her tree.

The next spring, new green shoots peeped up through the carpet of ash on the forest floor.

Top tips for telling

Start by evoking the love of the bird for her tree and home, and her determination not to let the fire destroy them. It can be fun to get the class to join in making the sounds of the fire, and the sounds of the stream, and the hiss as the water evaporates in the fire.

You can build drama as the fire gets closer and all the gods are laughing at the bird and predicting her death. Then, the character of the Eagle God is important. He is caring and touched by the bird's courage. This is a gentle and powerful tale for children to remember: an act of courage can have a big effect.

Ways to work with these stories

Teach one of the stories using HMSS

Deepen the story with activities such as these:

- Re-enact the story of *A Drop of Honey* emphasising the misunderstandings that cause the problems, and the attitude of the King.
- Role-play the King and his Minister discussing the various situations.
- Explore the dilemma of whether or not the King should intervene at each point. You could use a 'thought corridor' in which one child walks down between two lines of children: those in the line on the left whisper reasons why the King should intervene while those in the line on the right whisper reasons why he shouldn't. The child who walks down the centre, listening, has to decide what his decision would be and why.
- Role-play being a TV reporter commenting on the scene and the King's lack of action at each point.
- Re-enact a TV talk show in which the King, Minister and shepherd are interviewed. Who was to blame?
- Re-enact the story and stop it at various points and ask the characters why they are retaliating.
- Re-enact the story of *The Bird and the Forest Fire* emphasising the passion of the bird for her forest home.
- Use a thought corridor to explore the dilemma faced by the bird when she sees the forest is burning.
- Role-play an environmental expert and explain how one small change in the environment can have knock-on effects and lead to much bigger problems.
- Role-play being campaigners on behalf of an endangered species and explain to the listeners how their behaviour is causing a species to become extinct.

- Role-play a person who, faced with an impossible task, will not give up, e.g. watering a plant in a desert.

Explore the science that is directly linked to the stories:

- The actions of humans directly affect the environment for better or worse.
- Habitats involve complex relationships between plants and animals.
- Food chains including producers, predators and prey.

Explore the science that is indirectly linked to the story:

- grouping living things in a variety of ways, including with classification keys;
- how science helps us understand our world and predict the consequences of our actions;
- the way that bees make honey and how this activity is crucial for crop pollination.

Try these science activities:

CARING FOR THE ENVIRONMENT

1. Group living things

Collect a large selection of pictures of creatures or plastic toys. Encourage the children to look at them closely and decide which ones belong in a group together. Encourage them to sort using one feature, e.g. number of legs. Ask questions, e.g. Do birds and flies belong together? Do fish and seals belong together? Research the features that define a mammal, a reptile, an amphibian, a fish, an insect and other groups of invertebrates.

2. Use a key

A good place to start is to generate a set of questions that can be answered with a yes or no. Pick two items and generate a question for which the answer will be yes for one item and no for the other, e.g. Is it blue? You can use any items or even use pictures, e.g. cartoon characters, football players, shoes or flowers. Then give the children four items to sort and find a yes/no question that will separate the items into two groups. You now have the first question in a key that will sort the items. Keep asking yes/no questions until there is only one item left in each group. The set of questions can then be drawn out as a key. Look at biological keys and use them to sort animals. Choose four animals and write keys to sort them.

3. Collect minibeasts

Being mindful of the environment, encourage children to collect minibeasts from a variety of habitats and use keys to identify them. Notice their different features that enable them to live in their habitat. Always return the minibeasts to the places where they were found.

4. Design a minibeast home

Research the needs of a woodlouse. Make a choice chamber (a tray with a different habitat in each corner). Release woodlice into the tray and watch where they go. If you leave them for five minutes they'll choose an area that is cool, shady and damp and contains rotting wood. You could count how many end up in each area, repeat the process a few times and draw graphs of the data. Make sure you handle the beasts gently (paintbrushes are good for picking them up carefully) and put them back where you found them at the end of the lesson.

5. Food chains and ecosystems

Research the food chains in any habitats you can visit easily, such as a pond or hedgerow. Show that all food chains begin with green plants and the energy from the sun. It's fun to demonstrate this with a ball of string. Stand all the children in a circle and ask one child to hold the end of the string. Then zigzag the string across the circle and back, with each child holding onto a point along the length of string, making a web in the middle of the circle so that all are connected by strings. Then ask one child to pull their string down to the floor to represent the extinction of a species in the ecosystem. This string will pull down on all the other strings. If a child feels their string being pulled then they are 'extinct' too and must pull their string down to the floor, demonstrating the interconnectedness of an ecosystem. If one species becomes extinct, this can cause the extinction of others.

6. Environmental change and extinction

Choose an animal on the brink of extinction. Encourage the children to research and discover the reasons for the species' plight. In particular, try to link these reasons back to human behaviour. Relate this to the story of *The Drop of Honey*. Record as a fact file or on video as a TV report.

Explore other curriculum areas that link to the stories:

Reading

Read other stories with an environmental message, e.g. *The Lorax* by Dr Seuss (Random House, New York, 1971) or *The Man Who Planted Trees* by Jean Giono (Shambhala, Boston, 2000).

Window by Jeannie Baker (Walker Books, London, 2002) is a wordless book that shows the view from a window changing over a lifetime. It is a striking example of how humans change their landscape and a great book to study with this topic.

Fiction

Box up and write the story, or innovate using a different chain of events or in a different context where things escalate because no one intervenes.

Non-fiction

Discussion: 'Should we ban people from cutting down trees?'

Discussion: Discuss whether the King should have got involved.

Explanation: Write an explanation of how wasting energy leads to global warming and how this can lead to a particular species becoming extinct. You could link this to endangered species in your own local environment.

Persuasion: Write a letter to a large corporation to persuade them to take steps to protect the environment.

Persuasion: Be the King's Minister and persuade the King to call a truce to the war.

Persuasion: Persuade the rest of the school community to reduce, reuse and recycle.

Art

Sketch plants and animals that you have found in different habitats.

Study landscape artists. Compare the country landscapes of Constable with the industrial landscapes of Lowry.

History

Investigate the impact of human behaviour on wildlife. One good example is the discovery of Mauritius and how this led to the extinction of the dodo.

Investigate the impact of the industrial revolution, in the 19th century, on wildlife.

Life Skills

Discuss how our own actions can lead to unexpected and unwanted changes. Find ways to care for our environment.

Topic 3.6 HEALTH AND GERMS

Story: The Story of Edward Jenner

Topic: Healthy living, microbes that harm us, medicines and scientific testing
Plot type: Quest
Genre: Historical fiction

This is the true story of a doctor called Edward Jenner who, with the help of a brave young boy, managed to change the world by finding a vaccine for smallpox. He discovered, by listening to one of his patients, a milkmaid, that girls who caught cowpox were protected from getting smallpox. Jenner tested this theory and found that it worked.

The science content of the story illustrates how scientists work by testing out a theory. It also shows how some germs can harm us and that medicines can protect us and make us well.

The Story of Edward Jenner

Once ... there was a boy called Edward Jenner. He lived in a small village in the countryside. It was a quiet life. He went to school and he played with his friends. He watched the plants grow season by season. He watched the farmers in the fields and dreamed of what he might be when he grew up. A farmer? A miller? A baker?

And then, when Edward was still a child, he suddenly realised exactly what he wanted to do.

He had heard about an illness called smallpox. It was a killer disease. First, you got boils on your skin and you felt really sick. Then came the fever. Many people died. In some villages, hardly anyone survived the smallpox.

Edward decided to be a doctor, to try to help people who had smallpox and other such diseases.

He studied hard at school, then went off to university and became a doctor. He was a dedicated and careful doctor. Every day, people came to his surgery, and he helped them as much as he could, but always, at the back of his mind, he was pondering the problem of smallpox and how to find a cure.

One day, a milkmaid came to see him. She showed him her hands. She had some scabs. Edward knew at once that this was cowpox from the cows she milked. Cowpox was harmless. Cowpox didn't kill you.

'You'll be fine,' he told her. 'This is just cowpox. You've caught it from the cows you milk but it won't make you ill.'

'That's good,' said the milkmaid. 'I don't mind having cowpox because I won't get smallpox now. Milkmaids who get the cowpox never get smallpox. Every milkmaid knows that. That's why milkmaids have such lovely skin! Cowpox doesn't leave many scars.'

Edward's mind started racing. What if this was true? What if getting cowpox did stop you from getting smallpox? Edward wondered if this might be the cure he was looking for.

He decided to test it out scientifically.

Edward had a gardener who had a son called James. When Edward explained his plan the gardener agreed to let his boy help. James was perfect for the test because he had never had either kind of pox.

Edward took pus from the cowpox sores on the milkmaid's hand. He scratched James's hand and put the pus on the scratches so the cowpox would get into the boy's body.

After a few days, boils appeared on James's hands. Edward examined them. It was cowpox. Then, when James was better, Edward called for him to come and see him. It was time to try and infect him with a mild form of smallpox.

Edward looked at the boy and took a deep breath. 'I hope this works, James!'

'Mr Jenner, sir, it's alright. I trust you,' he said.

Edward smiled. What a brave young boy! He scratched James's hand and put a tiny amount of the mild form of smallpox on the scratches.

He watched James carefully every day, looking for signs of smallpox. He took his temperature. He checked his body for boils. He watched and he waited and he checked. The temperature went up a little but then returned to normal. He watched and he waited and he checked for three whole weeks ...

But the boy did not get ill!

Usually, with tiny amounts of this mild form of smallpox there was fever and small sores, but James showed no signs of anything like that.

James was fine. Edward was delighted. This could mean a cure for smallpox, his life's dream. It looked as if the milkmaid was right.

Edward Jenner tried the same process on twenty-three more people to check that it worked for everyone. No one got sick at all.

Now he was sure he had done it. He had the cure for smallpox! He had proved that if a person had had cowpox then they couldn't get smallpox.

He called his method a 'vaccine' after the Latin word for 'cow'. It took a while to convince people that his vaccine would work. The doctors in the city looked down on him because he was only a country doctor.

But eventually the smallpox vaccine was developed and used everywhere.

Smallpox has now been eradicated – there is no smallpox anywhere in the world. You will have had several vaccines in your life. They've protected you from other killer diseases. And you owe it all to Edward Jenner ... and he owes it to the milkmaid!

Top tips for telling

First, we need to show the determination and drive of Jenner to do something about smallpox. Then, when the milkmaid turns up, evoke his excitement at the possibility of finding a way to protect people. Finally, find a brave voice for the young boy who wants to help with the test in spite of the dangers.

This is a quest story with the whole world benefiting from Jenner's idea. Make sure that the effect on the world is emphasised in how you conclude the story and link it to your class's own health and well-being.

Topic 3.6 HEALTH AND GERMS

Alternative story: The Broad Street Pump

Topic: Healthy living, microbes that harm us, medicines and scientific testing
Plot type: Quest
Genre: Historical fiction

This is the true story of a doctor called Dr Snow who proved that people caught cholera by drinking contaminated water not by breathing bad air. He traced the source of an outbreak of cholera to one pump in Soho in London.

The science content of the story illustrates how scientists work by collecting evidence to prove a theory. It also shows how some germs can harm us.

The Broad Street Pump

Once ... in the middle of the 19th century, there was a terrible disease, which killed thousands of people all over Europe and beyond. The name of the disease was cholera. When people got sick they suffered from such severe vomiting and diarrhoea that their bodies lost too much water and they died. At that time, most people believed that the disease was spread by people breathing in bad air.

There were frequent cholera outbreaks in London. One doctor who lived there, Dr Snow, had a different idea. He thought that the disease was caused by drinking dirty water, but nobody took him seriously. He told the authorities that providing clean drinking water would stop the illness, but they didn't believe him.

In London at that time, there were no water pipes or sewer pipes. People collected water from hand pumps that sucked water up out of the ground, and threw their dirty sewage into the river or into holes in the ground by their houses.

Dr Snow decided to prove, scientifically, that dirty water was the cause of cholera. This is what he did.

When a cholera outbreak began in Broad Street in the Soho part of London, Dr Snow made a study.

First, he visited all the houses in the area and noted down in which homes people were sick with cholera, and in which they were healthy. Then he asked the people in every house, 'Where did you get your drinking water from?'

Some said they got water from the old hand pump in Broad Street. Others said they got water from different pumps further away.

After that, he made a map showing the houses where people were sick and the houses where they were healthy and mapped the pumps where they collected their water.

He found that all the people who were sick with cholera had drunk from the Broad Street pump. People who drank from the other pumps were fine.

Dr Snow went to the government with his map and argued his case with them.

'Look at my evidence,' he said. 'This proves, scientifically, that dirty water is the cause. Only people drinking from this pump are getting sick!'

The authorities agreed to take the handle off the Broad Street pump so it couldn't be used.

After that, Dr Snow went back to the homes to see what happened and within a few weeks there were no new cases of cholera. He also found a sewage pit right next to the Broad Street pump.

Dr Snow argued that his evidence proved that water, not air, was the cause of cholera.

A few years later a German scientist, Dr Koch, discovered the bacterium that causes cholera, a bacterium that is found in the toilet water of sick people.

As a result, the main cities of Europe began to build water pipes and sewers to keep drinking water clean and separate from waste water, and they are still used today.

Top tips for telling

This is a kind of quest story with Dr Snow as the man on a mission. First, he collects his evidence. Then he persuades the authorities to take the handle off the pump because his evidence is so strong. Then the people stop being sick. Finally, when the cause of cholera is understood, everyone agrees that water needs to be clean and water supply pipes are built. So, you need to develop Dr Snow's character as a determined man and the drama when he tries to persuade the authorities to do something.

Apart from that, the science is about research and mapping, so be really clear about the steps in the study and how it proves his point.

Ways to work with these stories

Teach one of the stories using HMSS

Deepen the story with activities such as these:

- Re-enact the story making sure the enormity of the risk Jenner is taking with James's health is emphasised.
- Role-play Jenner explaining to James and his father how he plans to test the theory.
- Explore the dilemma that James and his father face when they are asked to help Jenner. You could do this using a thought corridor (see p. 66) with reasons to help Jenner and reasons not to help him.
- Role-play a doctor visiting a family with smallpox or cholera.
- Re-enact the story of Dr Snow presenting his evidence and conclusions to the government.
- Role-play Dr Snow explaining to the local people why their pump is being dismantled.

Explore the science that is directly linked to the stories:

- microbes and how they can be harmful;
- staying healthy;
- medicines and drugs;
- how real discoveries are made and theories are tested.

Try these science activities:

HEALTH AND GERMS

1. Microbes that cause illness

In response to the story, find out about germs we encounter in our daily life which can make us ill and how we can stay healthy. Map and step the stages involved when germs are passed from person to person.

2. Microbes that cause decay

Find out how foods are preserved to keep them from rotting. Relate these methods back to the microbes, e.g. drying foods removes the water so microbes can't survive whereas freezing makes the food so cold that the microbes can't grow. Show how decay is part of natural recycling and is essential for the world. Investigate decay – you could leave tomatoes (cut or whole) in different places (in sealed containers) and observe the different rates of decay. You could put one in oil/salt/brine to show that it is preserved.

3. Microbes that help us

Find out why we eat 'live' yoghurt. Make bread dough using yeast. You could even grow a culture of live yeast. Find out how cheese is made.

4. Staying healthy – exercise

Investigate what happens to the heart rate when we rest and when we exercise. Relate this to the function of the heart and lungs and how exercise makes us healthier. Find out how our heart rate and breathing rate increase to supply the increased demand for oxygen from our muscles.

5. Testing a theory

Come up with a fun theory, linked to heart rate and exercise, to test e.g. 'I think that skipping is a harder exercise than running and raises the heart rate more'. Devise a fair test that can be repeated a few times to get valid results. Carry out the experiment and collect the data. If you need a simpler investigation that doesn't require you to take heart-rate measurements, you could try something like 'I think children with the longest legs can jump the furthest'. Emphasise that the study of science often involves collecting evidence to prove a theory.

6. Medicines and drugs

Invite a nurse or other medical expert into school to talk to the children about medicines that are safe when prescribed by a doctor and the dangers of putting any other chemicals in your body.

Explore other curriculum areas that link to the stories:

Reading

Read other stories of scientific discovery, e.g. *Germ Hunter: A Story about Louis Pasteur* by Elaine Marie Alphin (Lerner, Minneapolis, 2004).

Narrative

Box up and write the story.

Non-fiction

Information report: Give a presentation about the eradication of smallpox.

Information report: Write a report about a disease such as measles or malaria that is widespread today and how it can be treated.

Recount: Write a report on how you tested a theory such as 'I think people with longer legs can jump further'. Devise a test and collect data and recount the way you carried out your test.

Recount : Write a newspaper report about a scientific discovery.

Recount: Write a formal biography of Edward Jenner or Dr Snow.

Discussion: Debate whether James should take part in the experiment.

Persuasion: Write a letter from James persuading his father to let him take part in Jenner's experiment.

Persuasion: Write a letter from Dr Snow persuading the authorities to take the handle off the Broad Street pump.

Explanation: How washing our hands helps to keep us healthy.

History

Find out about other medical discoveries such as the discovery of penicillin by Alexander Fleming in 1928.

Life Skills

Jenner devoted his whole life to finding a cure for smallpox. To what cause might we choose to devote our lives? Consider our own life goals and ambitions by drawing our future selves or writing a letter from our future selves encouraging us to work hard today.

Topic 3.7 FLOWERING PLANTS

Story: Jack and the Giant's Peach

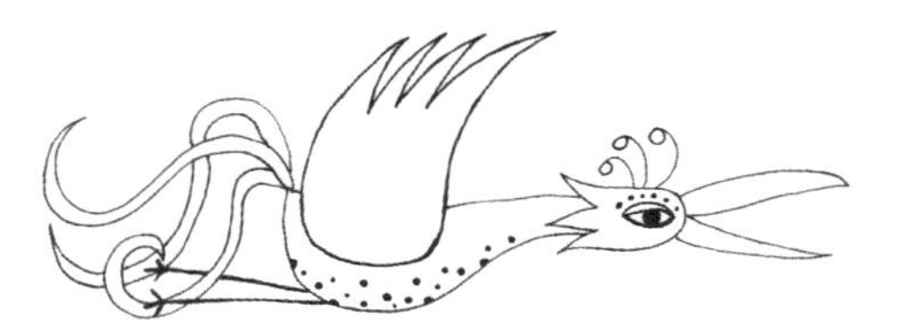

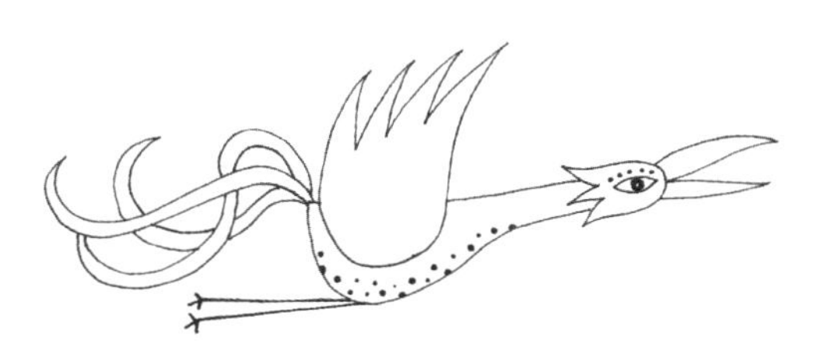

Topic: The life cycle of flowering plants, pollination and fruiting

Plot type: Quest

Genre: Wondertale

This is a variation on the traditional tale of 'Jack and the Beanstalk'. Jack climbs the beanstalk and reaches the giant's house and finds there a tree that grows giant peaches with golden stones. Jack knows a peach like that would solve all his problems. But the peach tree is still flowering. Jack has to keep returning until the fruit is ready so that he can steal a ripe one. He takes a fruit and falls into the hands of the giant. He only just escapes with his life and the peach.

The science content of the story includes all the stages in the life cycle of a flowering plant. The flowers on the fruit tree are pollinated, the seeds form and the fruits swell.

Jack and the Giant's Peach

Once ... there was a poor boy called Jack. Jack lived with his mother in a little cottage with Buttercup, their cow. One morning Jack's mother said,

'Jack, we've run out of food and we've run out of money. We'll have to sell the only thing we have. Take Buttercup to market and mind you get a good price for her.'

So, Jack went sadly off to market. He loved Buttercup but he had to do as his mother had said or there would be no breakfast in the morning. On the way to market, Jack met a mysterious old man. He was sitting at the edge of the old well and singing to himself.

'Wishes for sale,' the old man sang. 'Wishes for sale.'

Jack stopped and looked at the old well. If he'd had a coin he would have tossed it into the well and wished for plenty of food so that he could take Buttercup home with him again. The old man stopped singing and looked at Jack.

'Where are you off to with that sad look on your face, my boy?'

Jack told him how they had run out of food and run out of money, so he was off to sell Buttercup at the market.

'It just so happens that I need a cow,' said the old man. 'Sell me your cow, boy. I'll give you a good price.'

The old man held out a money pouch. In the pouch were beans.

'They're magic beans,' he explained. 'Magic beans that will make all your wishes come true.'

That sounded like a good deal to Jack. He had a lot of wishes and his mother did too. So he handed over the cow and went home.

Jack's mother was furious when she saw the beans. She was so angry that she threw the beans into the garden and sent Jack to bed without any supper. There wasn't any supper anyway.

In the morning, when Jack went into the garden, there was an enormous beanstalk growing there. It reached right up into the clouds. Without thinking twice, Jack shinned up the beanstalk and climbed up and up and up.

When he reached the top, he found himself in an unfamiliar world. Everything was giant sized, including the grass and the trees. He looked around. There were some enormous wooden gates and he decided to sneak underneath them to see what was inside. He seemed to have arrived in the garden of a giant house.

Suddenly, an enormous hand picked him up and he was face to face with a giant woman.

'A boy!' she boomed exasperatedly. 'If my ogre of a husband sees you, he'll eat you for dinner. His favourite dish is little boy pie. Run along home before he catches you. You haven't got long. He's just gone to check on his peaches with the golden stones.' And she flicked him out of the way like he was a pesky snail on her lettuces.

'Wow,' thought Jack. 'A peach like that would keep us going for years.'

'Excuse me,' he ventured. 'Did you say peaches with golden stones?'

'Yes, but there's no fruit yet; the flowers have only just opened. Now if you value your life, get going and don't come back.'

Jack headed off towards the beanstalk until the giant woman turned her back, then he changed direction and made for the peach tree. There he saw the giant. Hairy and ugly, he was watering the tree with a watering can that was as tall as Jack. Jack stayed well back and looked up at the tree. Huge white flowers the size of cartwheels were blooming all over the tree. Gigantic bees buzzed from flower to flower, carrying their cargo of sticky pollen from one flower to the next. Jack took a deep, deep breath of the sweet, sweet smell and decided to come back the following week.

The next week, Jack climbed back up the beanstalk. He made his way to the peach tree, hiding in the long grass behind the towering daisies in case the giant

should be nearby. The peach flowers had withered but under the flowers, where the seeds had formed, the fruit was beginning to swell. Jack took a deep, deep breath of the sweet, sweet smell and decided to come back the following week.

The next week, Jack climbed back up the beanstalk. He darted from giant flowerpot to tree trunk and made his way to the giant peach tree at the bottom of the garden. The flowers were gone and the fruit had swelled to enormous proportions. The peaches were hanging low, heavy and round and ripe. But ... under the peach tree, in an enormous deck chair, sat the giant. His eyes were shut and his breathing was heavy as he dozed.

Holding his breath, Jack began to climb the tree. The peaches smelt sweet and ripe and delicious. He found a beautiful blushing peach and began to hack at the stem with his penknife. It took ages. He should have brought the wood-chopping axe! Eventually, he could see that the peach was only holding on by a thread. As he cut through the very last part of the stalk, the peach dropped. It weighed a ton and as Jack grabbed at it, he slipped. Down and down he fell, closer and closer to the sleeping giant. Then, as if he'd heard the peach dropping, the giant shot out a hand and caught it – and caught Jack as well. Jack was trapped under his thumb. He wriggled and tugged in panic. The giant opened one eye. He drew the peach towards his enormous hairy face. When he saw Jack, wriggling and writhing under his thumb, he sat up in the chair, terrifyingly wide awake.

'A boy?' he roared. 'A boy stealing my precious peaches?

Steal a peach with a golden stone and I'll gobble you up, bone by bone!'

Quick as a flash, Jack pulled out his penknife and threw it in the eye of the giant. The giant roared in pain, dropped Jack and the peach and tried to pluck out the tiny knife with his enormous fumbling fingers. Jack, dragging the peach by the stalk, ran back to the beanstalk and threw himself down. He fell halfway down and clambered the rest as fast as he could. He ran to the woodshed and grabbed the wood-chopping axe and chopped frantically at the base of the beanstalk. He could hear the angry giant clambering down after him. Chop chop chop and the beanstalk began to tilt. Chop chop chop and the beanstalk came crashing down, carrying the giant with it. And that was the end of him.

Jack's mother was delighted with her peach and made a huge peach crumble for the whole village to share. The golden stone sold for enough money to buy Buttercup back and made them wealthy for the rest of their lives. So, Jack and his mother and Buttercup lived happily ever after.

Top tips for telling

The main thing to establish is the character of Jack (brave and in this story quite thoughtful) and the giant's wife (powerful but friendly) and then the menace of the giant as he wakes up. You need clear descriptions of the scale of everything in giant world and the excitement of action adventure as Jack climbs the tree, cuts the peach, stabs the giant and escapes down the beanstalk. For the science content, be clear about the details of the peach plants as you describe the flowering, bee pollination and fruit growth.

Topic 3.7 FLOWERING PLANTS

Alternative story: Emily's Bees

Topic: Flowering plants, pollination and the importance of bees in the process
Plot type: Overcoming the monster
Genre: General fiction

This is a story about a farmer's daughter who cares for the bees on the farm. When she moves away and the farm is sold, no one pays attention to the bees and they have to eat their own honey to survive. Eventually, they have to seek a new home. They find a place where the land is being farmed mindfully and it turns out to be Emily's new farm.

The science content in the story revolves around the bees being an integral part of the life cycle of flowering plants. The story demonstrates that the bees need a place to live and we should look after them.

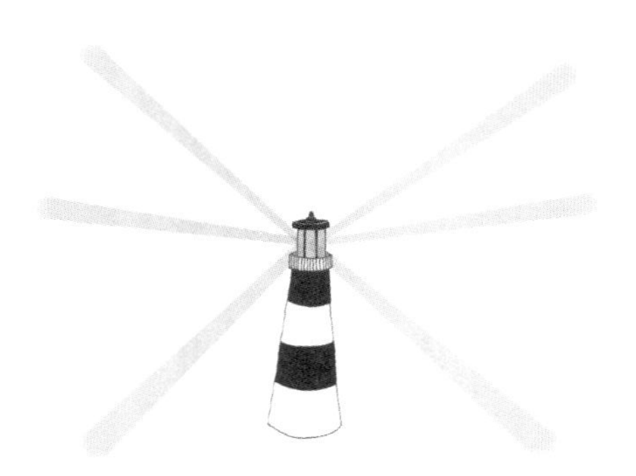

Emily's Bees

Once ... there was a farm. It was a patchwork of fields. There were three fields of pasture and hay, one full of flowers for the market, an orchard of apples and plums and pears, and a field of sunflowers. Between the fields were tall thick hedgerows, full of nests for birds and mice.

On the farm lived a farmer and his wife and their daughter Emily. Emily loved bees. She had her own beehive, which gave the family enough honey and wax for the whole year. Every day she'd go down to her beehive on the edge of her garden and sing to them:

Honey in the comb, honey on the comb
Bring the runny honey home
Share the pollen, as you feed
Making flowers into seeds

The bees loved the song and buzzed quietly along.

There were flowers everywhere from spring to autumn keeping the bees busy collecting nectar and pollen. Emily made sure there were always flowers growing nearby so that her bees could feed.

She loved watching them building their combs, laying their eggs and filling them with honey.

In this way the years passed and Emily grew older.

One day she came to the bees, sang her song to them and then said, 'Listen, bees. Soon I will be married, and I am going to move to my husband's farm. Mum and Dad are too old to farm alone so we have sold the farm. A new family is coming, but don't worry. This is your home and you can all stay here and help the new farmer just like you've helped me.'

The bees buzzed nervously.

A week later Emily was gone and the new farmer came.

The first thing he did was bulldoze the hedgerows and make one big field. The next thing he did was pull up all the orchard trees. Finally, he ploughed up the huge field, planted it with wheat and sprayed the field with chemicals every month to kill any insects that might damage the crop.

Now there were no flowers for the bees. And when some of the spray drifted in the wind towards the hives it made the bees feel ill and weak.

Soon they were hungry. They started to eat their own honey. When the honey was gone they decided to swarm. All the bees in the hive took off and flew away looking for a new home.

They searched and searched but everywhere they looked there were fields of wheat and corn. No flowers anywhere. No hedgerows. No fields of clover. And everywhere a horrible smell from the spray on the fields.

After three days flying they were desperate. Without flowers, they would soon starve to death.

Then they saw a patchwork of fields down the valley, with hedgerows, meadows and fields of flowers. They swarmed down into the garden on the farm and ... there was Emily, planting flowers in the garden.

'Bees!' she called. 'You are welcome. Look, I have a hive for you!'

The bees moved happily into Emily's new farm, and she sang to them again every morning:

Honey in the comb, honey on the comb
Bring the runny honey home
Share the pollen, as you feed
Making flowers into seeds

Top tips for telling

First, evoke the landscape of the farm and all the living creatures there. Then establish Emily's loving character and how much she loves bees. Then shift the mood with her big announcement. Evoke the panic of the bees as they fly around looking for a home and then their surprise and joy as they find Emily again.

Ways to work with these stories

Teach one of the stories using HMSS

Deepen the story with activities such as these:

- Explore the life cycle of a flowering plant as a dance or by stepping it.
- Re-enact the story with children portraying the peach tree in the different stages of growth and some playing the bees.
- Role-play the giant and his wife talking about the boy who keeps coming up the beanstalk and how they can protect their peaches.
- Role-play Jack reporting back to his mother about the progress of the peaches after each visit.
- Role-play a TV report on Jack as he makes his final descent with the peach and interview him about why it has taken so long to get one.
- Role-play Emily telling her husband about the bees she left behind and why she is worried about them.
- Role-play Emily persuading her husband not to use sprays on their farm.

Explore the science that is directly linked to the stories:

- The role of the flower in reproduction including pollination, seed formation and seed dispersal.

Explore the science that is indirectly linked to the stories:

- identify common wild and garden plants using keys;
- the requirements of plants for life and growth (air, light, water, nutrients from soil, and room to grow) and how they vary from plant to plant;
- the functions of the root, stem, leaves and flowers;
- how water is transported in plants.

Try these science activities:

FLOWERING PLANTS

1. The parts of a plant

Gather as many different plant types as you can: trees, bulbs, cacti, air plants, flowers, grasses, crops, etc. Have the children look closely at them to find the stem, roots, leaves, flowers, seeds, etc. on each. Talk about the function of each part.

Demonstrate the role of the roots as anchors by balancing a long cardboard tube on its end. It is very unstable. Tape four strings to the base of the tube, spread the strings wide apart on a desk and tape them down in that position. Now the tube is much more stable. It is made clear that a tall tree needs long roots to keep it upright!

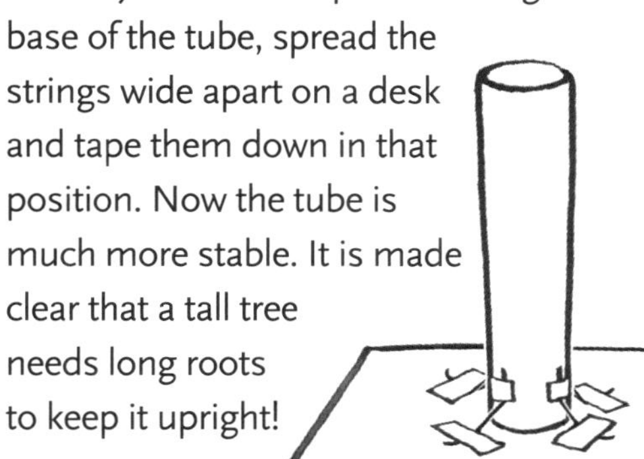

To show how water moves up the stem, put white carnations in water dyed with dark food colouring. Watch the colour move up the stem to the petals.

2. Photosynthesis

Research the leaf as the structure where food is made. Emphasise how important it is that plants can make their own food using the energy of the sun and how this process supports animal life. Observe plants that turn their leaves towards the sun. Move the plant so that its leaves face away and use a time-lapse photography app to film them moving back into the sun.

3. The parts of a flower

Dissect a range of different flowers. Compare the different shapes of the 'sticky stigma' and note that brightly coloured flowers attract insects to pollinate the flowers. Identify, draw and label the reproductive parts of the different flowers.

4. Pollination

Find out how the pollen reaches the egg. Look at wind-pollinated flowers and see how they differ from insect-pollinated flowers. Use paintbrushes to transfer pollen onto a flower.

5. Fruits and seeds

Dissect a range of fruits to see the seeds inside. See how the shrivelled flower can be found on one end of a fruit like an apple. Look at a branch of a tomato plant that has both flowers and fruit. See how the ovary swells to become the fruit. Hide pieces of different fruit inside cups so that the fruit cannot be seen but can be smelt. Identify fruits from their scent alone. Talk about why fruits are sweet and brightly coloured and how this helps with animal seed dispersal.

6. Seed dispersal challenge

Investigate wind dispersal. Look at a variety of seed heads and seeds from wind-dispersed plants. Give each child a seed and challenge the children to create a seed carrier that will catch the wind and take their seed far away from the mother plant. Give them foil, paper, tissue and other lightweight materials. Let them construct their own carrier around the seed. Test each one by dropping it from a height in front of a fan. Measure the distance it travels. The best ones float a long way and roll when they land.

Explore other curriculum areas that link to the stories:

Reading

Read other versions of *Jack and the Beanstalk*. See references in Chris Smith, *147 Traditional Stories for Primary School Children to Retell* (Hawthorn Press, Stroud, 2014).

Read *James and the Giant Peach* by Roald Dahl (Puffin Books, London, 2013).

Fiction

Box up and write the story, or innovate using different flowering plants.

Non-fiction

Information report: Give a presentation entitled 'Flowering Plants'.

Recount: Be Jack and tell your mother what you have seen up the beanstalk.

Recount: Set up a TV interview with Jack in which he recounts his story.

Discussion: Discuss whether Jack is a hero or a thief.

Persuasion: Be the old man and persuade Jack to sell the cow for beans.

Persuasion: Be Jack's mother and persuade Jack not to go up the beanstalk again.

Explanation: Write an explanation of how flowers are pollinated.

Explanation: Give a presentation on how magic beanstalks grow so tall.

Poetry

Write poems with a verse for each stage of the life cycle. You could do this for beans or peaches.

Art

Draw still life of plants, fruits, flowers and seeds.

History

Find out how Darwin and Mendel contributed to our understanding of plants and their evolution.

Chapter 4

Chemistry Stories

Topic 4.1 USES OF MATERIALS

Story: The Fairy Godmother's Day Off

Topic: Choosing the right material for the purpose and the properties of strength, waterproofing and flexibility of materials

Plot type: Comedy

Genre: Wondertale

This is a version of *Cinderella* in which the fairy godmother is having a day off and her assistant is sent to look after Cinderella. The assistant isn't very good at carriages. She tries a strong wooden box and a football before making a suitable carriage out of a pumpkin and getting Cinderella off to the ball just in time.

The science content in the story revolves around the need to pick the best material for the job. The carriage must be sturdy yet easy to cut and must also be waterproof. The wood is so strong that they can't cut a door into the wooden carriage. The football is too floppy once a door is cut, but the pumpkin can be cut and stays firm enough to make a strong and waterproof carriage to keep Cinderella dry in the rain.

The Fairy Godmother's Day Off

It was the day of the grand ball and everyone had already left in their very best clothes. Poor Cinderella was sitting by the fire, sobbing. She desperately wanted to go to the ball ... and meet the prince ... and dance until dawn ... but her wicked stepmother had said, 'No, look at you! You can't go to the palace. You must stay at home and make all the beds and plump all the pillows for when we get back.'

And so Cinderella had made all the beds and plumped all the pillows ... and now she was sitting by the fire, feeding a mouse with a crumb of cheese.

'I wish I could go to the ball,' she said to the mouse. 'I would love to meet the prince ... and dance until dawn.'

It just so happened that the mouse wasn't the only one listening. Someone else had heard her wish. It was a fairy godmother. Well, almost a fairy godmother. The real fairy godmother had taken the day off – which was most unfortunate for Cinderella. But the fairy godmother's apprentice, Dolores, had heard her wish. Dolores had passed all the tests on magical ballgowns and was very keen to tackle a real wish.

Dolores entered the room in a magical flash of light (she was good at that bit) and said, as grandly as she could, 'Cinderella, you shall go to the ball.'

Cinderella looked up. 'Oh, gosh, who are you ... where did you come from?'

'My name is Dolores. And I will be your fairy godmother this evening. Now stand up. Let's have a look at you.' Cinderella stood up and Dolores shook her head. 'No, no, no, this will never do. It needs sparkle. It needs sequins. It needs

pizzazz!' And with a flick of her wand she had transformed Cinderella's ragged dress into a beautiful ballgown (with matching glass slippers and coordinating handbag).

Dolores was pretty pleased with herself. Cinderella was delighted! She said a thousand thank-yous, hugged Dolores and skipped to the door.

It was pouring with rain and there were big muddy puddles all over the yard.

'No, no, no, this will never do. You'll need a carriage,' said Dolores.

Unfortunately, Dolores hadn't studied carriage conjuring yet. But she was determined to get Cinderella to the ball without asking for help. Out in the yard she could see an old cart. It didn't have a roof or a seat but there were wheels and ropes for a horse to pull it.

'Hmmm ... I don't know how to conjure a carriage but I can make things bigger. Have you got anything carriage shaped that I can work with? It needs to be waterproof in this awful weather.'

Cinderella, anxious to get going, grabbed the wooden spice box from the shelf. It was a lovely shiny dark wood. Dolores placed it on the cart.

'Yes,' she said. 'Very beautiful. That'll keep the rain off. Stand back.'

There was a flash and a good deal of sparks and the box enlarged into a beautiful wooden carriage. Cinderella clapped her hands and walked all around it.

'Where's the door?' she asked. The carriage had no door and no windows. The wood was solid.

'Bother,' said Dolores. 'I forgot about the door.' She waved her wand and the box shrank back to its normal size. 'I'll never be able to cut through this,' she said. 'Have you got anything else?'

Cinderella looked around the kitchen. 'How about the biscuit tin?' she asked.

'Perfect,' said Dolores. 'If you could just get me something to cut through this tin, I won't forget the doors this time!'

Cinderella looked doubtful. 'Would a breadknife do?' she asked.

'I'll give it a try!' said Dolores.

They sawed away with the breadknife but they just couldn't get the knife through it.

'We'd better try something else,' said Dolores. 'Have you got anything else I can work with? Remember it needs to be waterproof in this awful weather.'

Cinderella kept looking at the clock. It was getting late. She grabbed a football from by the back door. It would be waterproof and they could cut a door in it. She grabbed the big scissors and started cutting. POP! The football deflated completely.

'No, no, no, that will never do!' exclaimed Dolores. 'We need something stronger than that. It needs to be waterproof in this awful weather and it would help if we could cut you a door without it collapsing.'

Cinderella was skipping about from foot to foot. She was so anxious to get going. If it took any longer she might miss the ball altogether. And she really wanted to meet the Prince ... and dance until dawn. She felt she might burst into tears if she didn't go soon.

Suddenly, she saw the pumpkin in the larder. She grabbed the bread knife, took a deep calming breath and carefully cut off the top. She scooped out the seeds and cut a door and a window in the side. She even carved out a seat for herself. She put the top back on and handed it over to Dolores.

'Perfect,' said Dolores admiringly. 'Stand back!'

Dolores placed it on the cart in the yard and waved her wand. There was a flash and a good deal of sparks and the pumpkin grew and grew into a beautiful pumpkin-shaped, pumpkin-coloured, pumpkin carriage. It was waterproof and sturdy and it had a door and a window! Cinderella clapped her hands. With another flick of Dolores's wrist, the mouse grew in size until he was big enough to pull the carriage.

'Well,' said Dolores, 'it may not be golden but it'll get you to the palace and keep you dry. In you get. And remember to be back by midnight.'

Cinderella was in the pumpkin carriage and ready to go in no time at all.

'Thank you Dolores!' she called out as the carriage rolled away into the night. 'If I get to dance with the Prince tonight it will all be thanks to you.'

Top tips for telling

This is a comedy because we play with the original story in a funny way. It can be told with a silly and ironic tone, making your character voices quite silly and over the top. Dolores is the key character – you might make her enthusiastic but a bit foolish. Ham it up as Dolores gets it wrong and Cinderella gets increasingly disappointed. Then show their delight when the problem is solved. Make sure you describe the materials involved in the various attempts.

Ways to work with the story

Teach the story using HMSS

Deepen the story with activities such as these:

- Explore the properties of different materials by physically acting out how materials behave – stiff and hard like wood or bouncy like rubber.
- Re-enact the part of the story where Dolores tries to make a carriage from each object and act out a discussion between Cinderella and Dolores about whether or not it is a good object to use for the carriage.
- Role-play Dolores trying to make carriages out of all kinds of other objects and work out what makes it a good carriage or a poor carriage.
- Role-play Cinderella getting to the ball a little late and explaining why she was held up.

Explore the science that is directly linked to the story:

- naming everyday materials, including wood, plastic, glass, metal, water and rock;
- describe the properties of different materials;
- compare materials, including wood, metal, plastic, glass, brick, rock, paper and cardboard, to find the best one for the job;
- what happens to materials when you squash, bend, twist or stretch them.

Explore the science that is indirectly linked to the story:

- group materials on the basis of their simple physical properties.

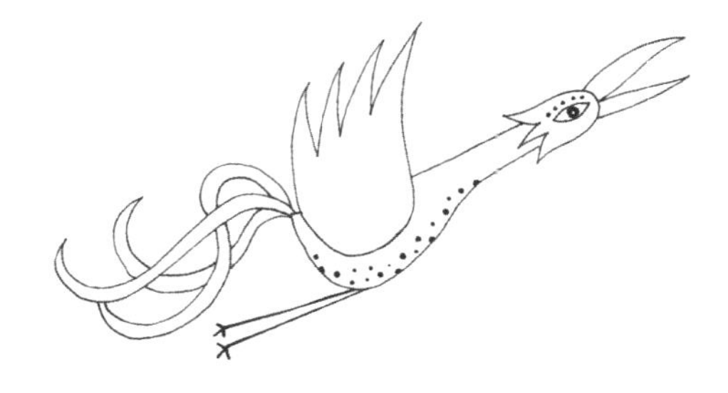

Try these science activities:

USES OF MATERIALS

1. Identify materials

Go on a treasure hunt. Find various materials (metal/plastic/wooden/fabric/clay/rock) and bring them (or photos) back to the classroom to feel and discuss. Make sure your hands are clean afterwards and that you don't pick up anything sharp. Collect sticks, peel off the bark to see the wood and smooth it with sandpaper. Watch videos of clay being made into plates and then fired and glazed. Make a class museum of different materials.

2. Sort materials

Show objects that might be part of a carriage: toy wheels, wooden bricks or planks, metal tubes, etc. Discuss what each is made of and why it is a good choice for the carriage. Ask the children to sort a selection of different objects into groups, e.g. round things. Encourage them to choose groups that are related to one another, e.g. different shapes – round, square, etc. Swap objects and have another go. Encourage them to use sensible groups based on properties. Record with photos or pictures.

3. Test for waterproofing

The carriage had to keep Cinderella dry in the rain. Test materials for waterproofing qualities. Discuss a fair way of testing them, e.g. put one drop of water onto a square of each material. Does it soak in? Try out ways to make things more waterproof, e.g. colouring thoroughly with wax crayon, coating with PVA glue, covering with cling film or covering with Sellotape. Some of these work better than others but all will make paper more resistant to soaking up water.

4. Cutting paper

In the story the wooden box and metal tin were too strong to cut but the football could be cut with scissors. Try cutting different types of paper. Which ones are hardest to cut with school scissors? Rank them in order from easiest to cut to hardest to cut. Investigate whether you can tear some more easily than others.

5. Test for strength

The wood was too strong to cut. Test the strength of different materials, e.g. plastic bags, toilet paper, newspaper and writing paper. Test each fairly by putting a 1 kg mass in a carrier bag. Lift it 10 cm off the floor using a 1 cm wide strip of each material– add weights and lift until the strip breaks. Identify the strongest ones. Relate the strength of the paper to the paper's purpose – toilet paper is designed to tear. Did you get the same results as when you cut them in the previous experiment? NB: Only do this activity on the floor, in a small group, under adult supervision, since you don't want 1 kg masses landing on your toes!

6. Fold and bend materials

The football was too floppy to make a good carriage. Investigate the flexibility of different materials. How many times can each material be folded?

Explore other curriculum areas that link to the story:

Reading

Read other versions of *Cinderella*. Roald Dahl's *Revolting Rhymes* (Puffin Books, London, 2013) has an alternative version. Read any traditional version of *The Three Little Pigs* and discuss the strength of each house. For traditional versions of both stories see *Jack and the Beanstalk: A Book Nursery of Nursery Stories* by Kathleen Lines (Oxford University Press, Oxford, 2013), or you can find *The Three Little Pigs* in *The Nursery Storybook* by Georgie Adams (Orion, London, 1996).

Fiction

Box up and write the story, or innovate using different objects/materials that make terrible or marvellous carriages.

Non-fiction

Instructions: Write a manual with instructions on how to make a carriage.

Instructions: Write instructions for Dolores on how to get Cinderella to the ball.

Explanation: Write an explanation of why the pumpkin was the most suitable object to enlarge into a carriage.

Persuasion: Be Cinderella persuading her stepmother to let her go to the ball.

Persuasion: Create an advert for the post of 'Fairy Godmother's Assistant'.

Recount: Be a TV royal reporter broadcasting live from the ball about the unusual carriage in which Cinderella arrives and interviewing her about how the carriage was made.

Music

Learn travelling songs, e.g. 'The Wheels on the Bus' from *The Wheels on the Bus* (BBC Audiobooks, 2007) or 'The Runaway Train' from *All Aboard the Runaway Train* (Jasmine Records, 1999).

Design/Technology

Make mini carriages for toys. This could be using building blocks, construction kits, or cardboard wheels and wooden dowel axles – you can do this really simply by taping two drinking straws to the underside of a box and putting the axles through the straws.

Life Skills

Invite an expert to talk about road safety to the children. Learn how to cross the road safely and have the children draw out chalk roads on the playground to practise their carriage-driving and road-crossing skills.

History

Learn about the history of transport. How did people travel before the motorcar was invented? When was the motorcar invented? How did your grandparents get to school?

Topic 4.2 CHANGING MATERIALS

Story: Death of a Pancake

Topic: Changing materials, melting and solidifying, cooking and changing shape

Plot type: Tragedy

Genre: Fable

This is a story about a pancake that jumps out of the frying pan and runs away. He is chased by lots of hungry people who want to eat him, but nobody can catch him. He finally stops to rest under a bush with a cunning fox, who offers him shelter and then gobbles him up when he is asleep.

The science content in the story includes the fact that some materials will melt when they are heated whereas other things will solidify. It includes examples such as butter melting, pancake batter cooking, iron softening and jam setting. It also includes other changes such as butter being churned from cream.

Death of a Pancake

Once there was a little old lady who lived in a little old cottage with her little old husband. One evening she decided to make pancakes for supper. First, she took down her mixing bowl and melted some butter over the fire. Next, she mixed the flour and milk and eggs and mixed it into a thick pancake mix. After that, she poured it into the pan and watched as it spread out and became a firm and delicious pancake.

The little old man came in and sniffed the air. 'Mmm, pancakes,' he said. 'They smell good.' He picked up one from the pan, folded it into quarters and started to eat it.

Then something amazing happened: the other pancake sat up in the pan, opened its eyes and screamed, 'Arrrgh! You are eating my friend!'

The pancake jumped down onto the floor and ran out of the house as fast as his pancake legs would carry him.

'Come back,' shouted the old woman, 'I'm hungry!' She ran after the pancake but she just couldn't catch him.

The pancake ran and ran and ran until he came to the milkmaid's house, where the milkmaid was churning cream to make butter. She smiled and said, 'Come in, little pancake, I have some cream left over and you will be very tasty with a big dollop of cream.'

'Arrrgh! You won't eat me!' cried the pancake and ran away as fast as his legs would carry him. The old woman and the milkmaid chased after him but they just couldn't catch him.

The pancake ran and ran and ran until he came to the jam-maker's and ran inside. The jam-maker was heating up blackberries in a big pot on the fire. The juice was thickening into a sticky jam.

'Come in, little pancake,' he said. 'I am making some fine jam and it will set soon. You will be very tasty with a big dollop of jam.'

'Arrrgh! You won't eat me!' cried the pancake and ran away as fast as his legs would carry him. The old woman, the milkmaid and the jam-maker chased after him but they just couldn't catch him.

The pancake ran and ran and ran until he came to the blacksmith's forge. The blacksmith was heating up an iron rod in the fire to make horseshoe nails. He pumped the bellows to make the flames hot until the metal was soft.

'Come in, little pancake,' he said. 'When I've finished making these nails I will put you on the fire till you are hot. You will be very tasty when you're all warmed up.'

'Arrrgh! You won't eat me!' cried the pancake and ran away as fast as his legs would carry him. The old woman, the milkmaid, the jam-maker and the blacksmith chased after him but they just couldn't catch him.

Soon, it started getting dark and the pancake looked around for a safe place to sleep for the night. He stopped by a bush and yawned. Next to the bush was Mr Fox. Mr Fox was hungry.

'Welcome,' said Mr Fox. 'Would you like to sleep here – safe under this bush?'

'Yes please,' said the pancake. 'Thank you. You are so nice. Everyone else wants to eat me!'

'No problem,' said the fox with a sly smile. 'I only want to help.'

The pancake closed his eyes and went to sleep, cosy and warm next to the smiling Mr Fox.

Once the pancake was fast asleep, the fox opened his mouth and ate him up. That was the end of that pancake, and this is the end of the story!

Top tips for telling

You've got two main things to do here as a storyteller. One is to evoke the drama of the 'life and death' chasing of the pancake. For that you need the fear of the pancake and the appetite of the other characters. The second is to evoke, clearly, in the imagination, the various objects that are changing from one state to another: pancake-cooking, butter-churning, jam-setting and nail-making. In each case, describe a little of what is seen as these things happen. This forms a basis for more investigation later on. It's a simple repeating story that you can tell with a clear rhythm so the children can join in as you tell it. Give it a slightly scary edge to make it more memorable.

Ways to work with the story

Teach the story using HMSS

Deepen the story with activities such as these:

- Explore in mime or dance the way different liquids and solids move.
- Re-enact the way that the different things are made: the pancake, butter, jam and nails
- Role-play the pancake telling Mr Fox about all the people who tried to eat him and Mr Fox listening patiently, licking his lips.
- Role-play one of the makers teaching an apprentice how to make the pancake/ butter/nails, describing what happens to the materials during the making.
- Role-play interviews with the makers about their methods as part of a TV documentary.
- Role-play a crime show hunting for the missing pancake with interviews and a call for information.
- Make up a poem about making pancakes (or butter, jam or nails) with all the senses involved.

Explore science that is directly linked to the story:

- Some materials can be squashed, twisted or bent.
- Some materials melt when heated and then solidify when they are cooled.
- Some materials change permanently when they are heated.
- Some materials change in other ways, e.g. churning.

Explore science that is indirectly linked to the story:

- describing materials
- comparing materials

Try these science activities:

CHANGING MATERIALS

1. Explore materials

Allow the children to feel and manipulate as many different materials as possible. Talk about how the materials feel and what they do. Sand trays and water trays are a great place to start. Then you can add trays of pasta or rice, or bowls of jelly. Our favourite substance to explore is gloop! You make it from cornflour and water. You'll need almost as much water as cornflour. When you get the mixture just right, it is a very thick sticky liquid that behaves like both a solid and a liquid. It is a non-Newtonian fluid, which means that it behaves in an unusual way. It is also very messy! Ask questions such as: Can you pour it? Can you bend it? Is it soft? Can you stretch it? Drop a weight onto it and observe it bouncing before it begins to slowly sink.

2. Change the shape of materials

Investigate bending and folding and squashing. Try moulding with playdough and clay. Is it easier to squash one? Drop balls of the clay or dough. Notice that the shape changes. Investigate what happens to balls of dough dropped from different heights.

3. Melt materials

Our favourite things to melt are ice lollies. You could even race a few different types to see which melts first. Also melt other everyday materials and show that they will reform when cooled. Wax crayons can be melted, by an adult, with a hairdryer. If you tape them to a canvas so that the melted wax drips down, it makes beautiful art but is rather splashy so stand well back. Butter, ice and candlewax are also easy to melt and re-

harden but, as they get hot, this activity will have to be carefully supervised by an adult.

4. Heat materials

Heat eggs, sugar, jelly and other household ingredients to see how they change. You can put them in a foil container over a single candle flame or use a pan on a stove or even a plate in a microwave. Sugar is best done in a pan. After a while you will see it begin to melt. Marshmallows are good fun in the microwave, since they expand in a dramatic fashion before collapsing in on themselves. All this must be done by an adult, since the ingredients will get hot. Observe how each ingredient changes.

5. Make pancakes

There are lots of recipes to choose from and you may want your pancake to be more like flatbread so you can have the experience of rolling it out. Whatever you choose to do, let the children feel the ingredients so that they get a hands-on sense of how mixing and heating the ingredients changes them.

6. Make butter

This can be done by hand or in a food mixer. Whip cream until it separates from the buttermilk and then drain the buttermilk away.

Explore other curriculum areas that link to the story:

Reading

Read other similar stories like *The Gingerbread Man*. There is a nice version in *The Nursery Storybook* by Georgie Adams (Orion, London, 1996).

Fiction

Box up and write the story, or innovate with different baked foods or things that involve a change in the material (cakes, biscuits, smoothies, cheese, road tarring, wall painting).

Religious Education

Find out about Shrove Tuesday and Mardi Gras and the origins of this tradition in the celebration of Lent.

Non-fiction

Instructions: Write instructions for a recipe book on how to make a pancake/jam/nail/butter.

Recount: Write a newspaper report about the pancake running away.

Recount: Be the pancake and tell Mr Fox about all the people who chased you.

Recount: Be the old woman and tell the old man how you tried to catch the pancake and what happened on the way.

Persuasion: Write an advert for the butter/pancakes you have made.

Art

Make models from clay – you could make your own runaway pancake. Leave them to dry and see how the material changes.

Physical Education

Explore the terms we are using in science lessons in gym lessons, e.g. 'bending', 'stretching' and 'rolling'.

Design/Technology and Maths

Measure the weight and volume of ingredients and make pancakes. You could even design your own decorations to give each pancake legs, arms and a face.

Topic 4.3 ROCKS AND FOSSILS

Story: The Fossil Woman

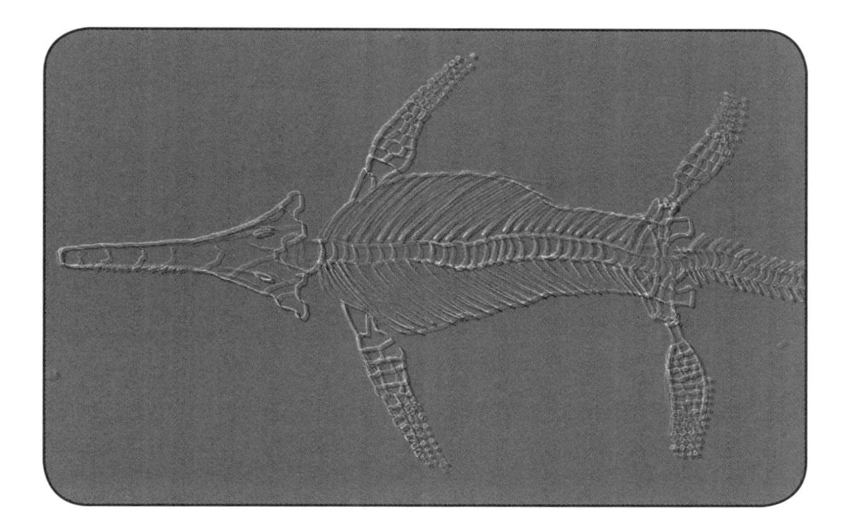

Topic: Rocks and fossils
Plot type: Rags to riches
Genre: Historical fiction

This is a true story about a poor family who make a few extra pennies by selling fossils and seashells to the visitors to their home town of Lyme Regis. When her father dies, Mary, the eldest daughter, has to work hard to bring in extra money to support the family. One day, she finds a huge skeleton buried in the cliff face. She chips it out of the rock with great care and sells it to the local lord of the manor. It was the first large prehistoric reptile to be found in Britain. It earned Mary enough money to buy a house for her family and fame as a palaeontologist.

The science content in the story includes the facts that skeletons buried in sediment, which hardens over time, will fossilise over time and these fossils tell us about life on Earth millions of years ago.

The Fossil Woman

If you go down to the Natural History Museum in London and walk through the tall oak doors, you will come to a room full of skeletons of dinosaurs and other large reptiles from all over the world. They are the remains of animals that lived millions of years ago.

One of the skeletons is from an ichthyosaur – which means 'fish lizard' – a prehistoric reptile as big as a bus. This was the first large prehistoric reptile ever to have been found in Britain. It was found two hundred years ago although it's millions of years old. It helped us learn all about our past. How do you think it got there? I'll tell you ...

Once ... down by the sea, in the town of Lyme Regis, there was a family who lived in a little cottage on the beach: a father and mother and their little baby, Mary.

Every day, the father would go off to work as a carpenter, sawing and hammering and measuring and gluing, making shelves and cabinets and all sorts of things out of wood. He took just a few pennies home every day.

One day, when Mary was under an elm tree, being looked after by three friends of the family, the tree was struck by lightning. The tree crashed down and all three women were killed. Their friends rushed over to help and when they picked up little Mary they were amazed to see that she was still alive.

Everyone was amazed. 'It's a miracle!' said one. 'She looks wide-awake and so happy.'

Mary grew up to be an energetic and curious girl whose favourite thing in the entire world was to go walking with her father on Sundays, collecting fossils by the beach. There were snakestones and devil's fingers – all sorts of fossils!

People in Mary's time didn't understand how fossils were made. We now know that fossils form underground when living things are buried in sediment. Over millions of years, as the soil or sand around them turns to rock, their shape is preserved. Sometimes the bones of an animal dissolve, leaving a space that is filled by minerals, so the shape of the bones can be seen. Sometimes the actual body parts are preserved within the rock. These fossils fascinated people, as they painted a picture about how life was in the distant past.

The family would find little fossils, old shells and little creatures from the sea and Mary would sell the shells and fossils to visitors on the sea front, calling out, 'She sells sea shells – see who sells sea shells!'

By now, Mary had many brothers and sisters. Times were hard and the family often went hungry. Some of Mary's brothers and sisters died from cold and hunger. So, the few pennies that Mary made really helped.

Then, when Mary was eleven, her father died of a chest infection and Mary, as the eldest child, had to earn money for the family. Without money the family would starve. Every day, Mary went along the beach looking for more fossils to sell.

One day she found what looked like a huge skull with eye sockets and teeth looking out from the rock face. She had never seen such a huge fossil before.

Mary took her hammer and started chipping, chip chip chip chip around the skull. Little by little the skull was revealed. After one month she could see the whole skull: it was longer than Mary was tall! Enormous! After three months, with a little help from friends, she had chipped out the whole thing. It was a huge, complete skeleton.

Mary was delighted. She sold the fossil to the local lord of the manor for twenty-five pounds! It was more money than Mary had ever dreamed of, enough to buy a new house with a big glass window, which became the Lyme Regis fossil shop. Her business grew and from that day her family never went hungry.

In this way, Mary became a well-known fossil collector. People came from all over the world to see and buy her fossils. She read and studied and hunted for fossils, selling them to museums all around the country, so others could learn about them too. Known all over the world as a fossil expert, Mary loved finding out about how life had been in the past.

In recognition of her work, the Royal Geological Society of London gave Mary a small annual income so she could keep collecting fossils. By the end of her life, she had discovered many fossils hidden in the rocks by the sea, which have helped us understand the prehistoric past. They can be found today in museums all over the world.

Mary was one of the world's first palaeontologists: she dedicated her life to learning about animals and plants that lived in ancient times.

Top tips for telling

There's not much dialogue in this story, but you can evoke the character of Mary with description through the voice of the narrator. This is a rags-to-riches story so make sure you show how poor her family was in the beginning and how much the money mattered once her father had died. Then there is the drama of the big find: make sure you show how amazing that moment was for Mary, and all the good things that happened as a result.

Ways to work with the story

Teach the story using HMSS

Deepen the story with activities such as these:

- Explore the process of fossil formation by demonstrating rock forming layer by layer with animal remains trapped within it. You could use sand and soil in a clear plastic box to build layers of sediment over a plastic toy animal until it is completely covered. You'll be able to see the layers through the sides of the box.

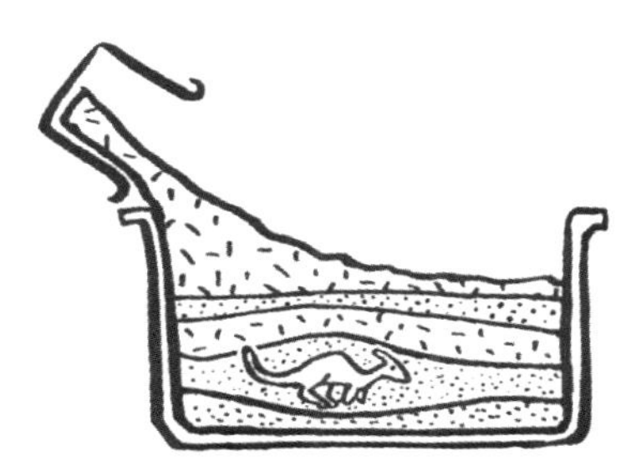

Get the children to make their own layers and encourage them to explain what is happening. There are fossil sets you can buy that enable children to chip away at 'rock' to find the fossils inside. You could use these as a prompt for explaining fossil formation too.

- Re-enact Mary discovering the skeleton as a drama documentary, with a commentator describing what she is doing and asking questions about what is going on.
- Role-play Mary giving a talk at the Royal Geological Society about the discovery of the ichthyosaur at Lyme Regis.
- Role-play geologists explaining how the different types of rock are formed and why you often find fossils in sedimentary rock.
- Role-play Mary in her fossil shop telling the visitors about the fossils.

Explore the science that is directly linked to the story:

- how fossils are formed when things that have lived are trapped within rock;
- what the fossilised remains can tell us about the animal when it was alive.

Explore the science that is indirectly linked to the story:

- the different types of soils and how they are formed;
- the different types of rock and how they are formed.

Try these science activities:

ROCKS AND FOSSILS

1. Sedimentary rocks

Find out how sedimentary rocks are formed. Examine sedimentary rocks closely with a hand lens. Feel the texture and relate this to the materials that made the rock. Watch videos about how such rocks are formed. Common examples are limestone, chalk, sandstone and shale.

2. Igneous rocks

Try making a bicarbonate of soda volcano as a fun way to start the lesson. Make from clay a hollow volcano shape with a vent at the top. Put a teaspoonful of bicarbonate of soda inside. Pour in vinegar and red food colouring and watch the volcano erupt! Research the different types of igneous rocks. Watch videos of volcanoes erupting and handle volcanic rocks to see how they differ from the sedimentary ones in their appearance. Granite and basalt are good examples of igneous rocks.

3. Metamorphic rocks

Find out how metamorphic rocks are formed. Examine metamorphic rocks closely with a hand lens. Watch videos about how they are formed. Slate and marble are good examples.

4. Testing rocks

Rocks vary greatly. Some are soft and porous like chalk whereas others are hard and impermeable. To test for hardness, you can scratch them first with fingernails and then with a steel nail. The softest will be scratched by both and the hardest by neither. Look at the Mohs scale of hardness to see which rocks are really hard. Test for permeability by putting drops of water on all the rock samples and see which soak in. You can also drop chalk in water and watch the air bubbles stream out – it can make quite a noise.

5. Fossils

A good way to study fossils is to visit your local natural history or geological collections or invite them to visit you. Observe fossils and draw them to become familiar with the common kinds.

6. Soils and soil formation

Put a few tablespoons of local soil into a beaker of water and mix well. Leave the mixture to settle and see the layers form – particles of similar size will settle into different layers showing that the soil has many components. Observe local soils on white paper with a hand lens or hand-held microscope. Test different types of soils to see their different properties. Test the permeability of the soils by putting equal quantities of soil into two funnels and pouring 100 ml of water through them, timing to see which drains first. (Use a plug of cotton wool in the neck of the funnel to stop the soil falling right through.) You could try rolling the soil into balls – clay-based soils will roll but sand-based soils will crumble.

Set up a wormery. If you fill it with distinct layers of sand and soil, the children can clearly see the role that worms play in mixing up the materials and aerating the soil for the plants.

7. Dinosaurs and extinction

Find out about ichthyosaurs and dinosaurs. Find out what it means to be extinct and about extinct species that we know about from the fossil record.

Explore other curriculum areas that link to the story:

Reading

Read about dinosaurs in non-fiction texts such as *The Big Book of Dinosaurs* (Dorling Kindersley, London, 1994).

Fiction

Create stories about the discovery of ancient creatures and how the discovery changes the life of the finder.

Narrative

Box up and write the story.

Non-fiction

Instructions: Write instructions on how to excavate a fossil.

Explanation: Explain to the class how fossils are formed.

Explanation: Write an explanation entitled 'Why Did the Dinosaurs Become Extinct?'

Information report: Write a poster about Tyrannosaurus rex.

Persuasion: Write a letter from Mary persuading her mum to let her be a fossil collector.

Persuasion: Write a fossil shop advert.

Recount: Be Mary as she tells her own story in her diary.

Recount: Be the dinosaur and tell your story.

Recount: Write a newspaper report on the day of the fossil discovery.

History

Create a timeline showing both the time when prehistoric reptiles such as ichthyosaurs and dinosaurs were alive and the time when the fossil was discovered by Mary Anning, as well as the present day.

Find out about archaeologists and how their work informs our understanding of what life was like in the past.

Study Pompeii and what the eruption of Vesuvius did to the town and how the preservation of the town has informed our understanding of the past.

Geography

Research volcanoes all over the world that are still active. Find out how various rocks are formed when lava cools. Obsidian is an interesting volcanic rock to research. Find out how the volcanoes can change the landscape of their area.

Art

Create a miniature dinosaur world! Find out about plant life in the prehistoric world and make plants and animals for a 3D display of dinosaurs and other prehistoric creatures in the world they inhabited.

Topic 4.4 PROPERTIES OF MATERIALS

Story: The Horses of Troy

Topic: Properties of materials including transparency, ability to absorb water, thermal conductance, flexibility and strength

Plot type: Overcoming the monster

Genre: Legend

This is a story about an inventor, imprisoned for desertion, who is offered his freedom in return for making Odysseus a horse to end the siege of Troy. He has only a few days to complete the task. He tries building a horse out of lots of different materials but they all prove to be problematic in some way. Eventually, he settles upon wood as the most suitable material and wins his freedom.

The science content in the story is all about choosing the best material for the purpose. There are five horses made from different materials which can be included as required. The first is made of paper that is lightweight but absorbs water. The second is made of metal and demonstrates the property of thermal conductance. The last horse is made of wood and demonstrates how wood is lightweight, strong, waterproof and a good thermal insulator and thus good for the task. The glass eyes can link to the topic of transparency.

We also offer a longer version of the story which includes a clay and a stone horse, linking to the properties of thermal insulation and strength, respectively.

The Horses of Troy

Have you ever heard the story of the wooden horse of Troy? Paris the prince of Troy stole the beautiful Queen Helen of Greece away. He locked her up inside the walled city of Troy. The Greek army came to rescue her, of course. There was a siege. The Greek army sat waiting outside the gates of Troy because they couldn't get in. They tried to starve the Trojans out but the Trojans had supplies stashed inside. They were fine. The Greeks were never going to get in that way.

So they made a huge wooden horse, left it on the beach, pretended to give up on the fight and sailed away. The Trojans were curious and came out to look at the horse. It seemed safe and it was rather impressive, so they took it inside the gates as a victory trophy. But in the night some Greeks, who had been hiding inside the horse, opened the gates of the city and the Greek army stormed in while the Trojans slept.

But what of the wooden horse – how did that come to be? Maybe this is what happened ...

The Greek army had laid siege to Troy for ten long years. In the Greek camp, there was an inventor called Archippos. He liked designing things, but he was useless in a battle. He couldn't fight. He was so afraid that he ran away from the fighting but he was caught and sentenced to death for desertion.

One night, wily Odysseus came to visit him in prison.

'Archippos,' he said, 'I need to smuggle some men into the city of Troy. I want to hide them in a huge horse that looks like a fine statue. We will sail away and the Trojans will think they are safe and they will take the horse into their city as a souvenir. Make one for me and I will spare your life. You have three days!'

Archippos was hopeful. Maybe now he could escape this endless war and get back home to his beautiful wife. Odysseus took him to a tent full of tools and materials and set two guards at the door. First, Archippos heated up some sand and melted it into two almost transparent eyes for the horse. If you pressed your eye right up against them you could see out. The soldiers inside would be able to see when it was dark and they could jump out and surprise the Trojans. He put the eyes aside to cool and got on with the body.

He needed something he could mould ... Papyrus. So, behind the tent, he began to mould the horse out of papyrus mulch. He squidged and he squeezed it with his hands as he shaped it into two fine hooves. And as he worked he sang:

Giddy up, giddy up, I want to go home
So much to do, so little time
I hope this horse will carry me home
To the land and the lady and the love that's mine

He worked until nightfall, and then left the finished horse outside to dry. That night it rained. When Archippos awoke in the morning the horse had turned to mush.

'O Hades, this is never going to work!' he cried. 'Odysseus will hang me for this. I'll have to try something else. I need something strong and waterproof! I know ... metal!'

So he began to shape a horse out of metal panels. He hammered the metal into four fine legs. And as he worked he sang:

Giddy up, giddy up, I want to go home
So much to do, so little time
I hope this horse will carry me home
To the land and the lady and the love that's mine

The next morning dawned hot and sunny. When Archippos put his hand on the metal it burned.

'OW! – O Hades, this is never going to work! The soldiers would fry in a metal horse under this hot sun. Odysseus will chop off my head. I'll have to try something else. I need something strong and waterproof that doesn't get too hot.

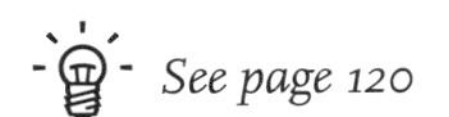
See page 120

'I know ... Wood!'

He was in a real hurry now. Time was getting short. He was so tired and he ached to go home. This last horse had better work. So he set to work carving away. He hammered and sawed. He chipped and he chiselled. The woodchips flew all over the place. And as he worked he sang:

Giddy up, giddy up, I want to go home
So much to do, so little time
I hope this horse will carry me home
To the land and the lady and the love that's mine

At last the wooden horse was ready. Archippos fetched the transparent eyes he had made and slotted them into place. It was a beautiful horse. Archippos put away the tools and hoped that it would be good enough to win his freedom.

The next morning Odysseus came. It was time ...

Odysseus looked the horse up and down. He ran his hands over the smooth wood and he tried the hatch in the horse's belly. He climbed inside and looked out through the glass eye windows. Archippos stood at the bottom of the ladder waiting. He searched Odysseus's face for any sign of approval but he could see none.

'Archippos, we made a deal,' said Odysseus. 'A horse in return for your life ... This is a fine horse and I am a fair man. You are free to go. My guards will not stop you. There is a ship at the dock leaving for home tonight.'

Archippos started running and never looked back.

Years later, on an island, far away from Troy, a man sat beneath a shady tree listening to a storyteller. On his lap sat a small boy. They listened as the storyteller told how clever Odysseus had rescued Queen Helen and defeated Troy by tricking the Greeks with a wooden gift horse, full of soldiers. The man began to bounce the boy on his knee and to sing to his son:

Giddy up, giddy up, I want to go home
So much to do, so little time
Maybe this horse will carry me back
To the land and the lady and the love that's mine

And as he cradled his son in his arms he whispered, 'One day, I'll tell you all about it.'

If you like you can extend the story to include a clay and a stone horse, maybe by inserting the following sections at point in the story or you could get your class to make up their own narrative for these and other materials.

'I know ... clay!'

So Archippos began moulding the horse out of clay. He made four legs and a body. And as he worked he sang:

Giddy up, giddy up, I want to go home
So much to do, so little time
I hope this horse will carry me home
To the land and the lady and the love that's mine

The next morning, he came back with a fine horse's tail, made of real horsehair, to decorate the clay horse. He climbed up and tapped a nail into the horse's bottom to hold the tail. And CRACK! The clay horse broke into a million tiny pieces and clattered to the floor.

'O Hades, this is never going to work! Odysseus is going to pierce my heart with an arrow! I'd better try something else. I need something strong and waterproof that doesn't get too hot and won't shatter when I hit it.

'I know .... Stone.'

Archippos called for some stone and some stonemasons to help him and they all chipped away at the stone with their chisels, carving and shaping until they were up to the neck. And as Archippos worked he sang:

Giddy up, giddy up, I want to go home
So much to do, so little time
I hope this horse will carry me home
To the land and the lady and the love that's mine

The next morning, he had made a beautiful carved stone horse's head. He'd been up all night carving it. He put it in a hoist to lift it up onto the horse's neck.

'Help me roll the horse to the hoist,' he called out and the servants came running to help.

Heave! Heave! They pushed and they pushed but they couldn't get the horse to move an inch.

'O Hades, this is never going to work! The horse is too heavy to move. Odysseus will have my head on a stake! I need something strong and waterproof, cool and shatterproof, and light enough to be able to push it along.'

Top tips for telling

First, decide if you are going to use the three-horse or the five-horse story. With three horses the inventor gets three days to invent; with five horses he gets five days.

Get the setting clear in your mind. This is the siege of Troy, and the Greeks are living mainly in tents and ships on the beach.

Now establish the character of the inventor: he is clever and frightened and wants to go home. Find a simple tune to go with the song: two or three notes are enough. You can sing it like a sea shanty.

As you tell, emphasise the descriptions of the horses: picture each in your mind before you start telling to make it as vivid as possible. Evoke the cycles of enthusiasm and disappointment, and finally relief as the problem is solved.

Ways to work with the story

Teach the story using HMSS

Deepen the story with activities such as these:

- Re-enact the building of each horse using suitable tools and materials.
- Role-play being a wise old man who advises Archippos about the different materials and explains why wood would make the best horse.
- Make up a poem about each material and the properties it has.
- Role-play a news reporter interviewing Archippos every morning as he starts building each horse.
- Role-play being market traders advertising your paper/metal/wood to Archippos as he goes shopping to buy what he needs for the horse.

Explore the science that is directly linked to the story:

- properties of everyday materials, e.g. hardness, solubility, transparency, conductivity (electrical and thermal);
- sorting materials based on their properties;
- choosing the right materials for the job.

Explore the science that is indirectly linked to the story:

- Different soils have different properties (if you include the clay horse).
- Different rocks have different properties (if you include the stone horse).

Try these science activities:

PROPERTIES OF MATERIALS

1. Testing strength

Measure the strength of different kinds of paper – lift bags of 100 g masses with strips of paper to see which is the strongest and carries most before breaking (see p. 99). You could try sugar paper, photocopier paper, newspaper and kitchen towel. Relate this to the fact that Archippos might have chosen the strongest papyrus. Encourage the children to make the test fair and remember they need to work on the floor with adult supervision. Once you have collected the results, draw bar charts to show which paper was the strongest.

2. Testing absorbency

Compare the absorbency/waterproofing of different paper – measure how far water will soak up a strip of paper that has one end dipped in some water coloured with food colouring (this makes it easier to see). Talk about how to make this test fair, e.g. same length of paper, same time left in the water, etc. Archippos might have chosen a papyrus that absorbs water well to make his papyrus horse.

3. Testing thermal insulation

Fill empty drinks cans with water. You could use hot water to investigate the speed at which it cools or use cold water and investigate the speed at which it warms. Wrap each can in a different material to identify the best thermal insulator. Talk about how to make this test fair, e.g. one layer of material, water at the same temperature, etc. Relate this to Archippos needing a material that is a good thermal insulator for the horse.

4. Testing transparency

Compare the transparency of different materials by looking through them to see how much detail is visible. It is possible to read through the most transparent materials. Archippos needs transparent eyes on the horse to act as windows.

5. Testing soils (if you include the clay horse)

Compare different soils. Try to find a soil that is mostly clay and see how it behaves compared with sandy soils when you roll it and squash it. Investigate the properties that each soil has. Make pots from clay and leave them to dry. If you are able to fire the pots then you could compare the properties of fired clay and air-dried clay.

6. Testing rocks (if you include the stone horse)

Investigate the different types of rock that Archippos might have chosen. Try to find a type of stone that can be chipped or carved. Investigate the other properties this stone has. Look at the rocks in your local area and learn to recognise and name them. Research the types of stone that were commonly used in ancient Greek buildings and statues.

Explore other curriculum areas that link to the story:

Reading

Read other Odysseus stories and different versions of *The Horse of Troy*. *The Odyssey* by Rosemary Sutcliff (Frances Lincoln, London, 2014) is a lovely version of the story.

Fiction

Box up and write the story, or innovate using other materials (bricks, leaves, ice, animal skins).

Non-fiction

Recount: Prepare a TV interview with Archippos back home or a newspaper report on the wooden horse of Troy.

Instructions: Give instructions on how to make a wooden horse (this could be a YouTube clip or manual).

Instructions: Write instructions on how to rescue a captured queen (or king).

Discussion: Discuss which is the best material to make the horse (pros and cons of paper, stone and wood).

Persuasion: Create an advert or a sales pitch to promote the model you made in Design/Technology.

Explanation: Write an explanation entitled 'Why Wood Is the Most Suitable Material for the Horse at Troy'.

History

Find out about the ancient Greeks.

Study the influence of the Greeks on the Western world.

Art

Look at photos of ancient Greek pottery and see if you can recognise the stories that are told in the decorations upon them.

Design/Technology

Make recycled paper and investigate how papyrus was made.

Investigate ways to make paper stronger: fold it, roll it into columns, layer it, etc.

Design an egg carton to protect an egg when dropped in the carton from a height. You could base your designs on eggboxes or Easter egg packaging, since both provide protection. The choice of materials could be limited to types of paper so that children have to create carton shapes that will absorb the impact. Or you could allow the children to suggest their own choice of materials, so they get the chance to suggest materials that will, themselves, absorb the impact regardless of the shape of the carton.

Topic 4.5 THE WATER CYCLE

Story: The Children of the Water God

Topic: The water cycle, evaporation and condensation
Plot type: Rebirth
Genre: Creation myth

This is a story about the two children of the Water God. The Water God owns all the water in the world and refuses to share it with the ever increasing human population. When the thirsty people are forced to steal enough water to survive, the Water God sends his two children to guard his water.

Warmhearted Eva and her cooler brother Con find a way to get water to the people by turning the seas into clouds, which rain on the land. When the Water God finds out, he is angry but his children show him that all the water returns to him in the end, so he relents and gives his blessing.

The science in the story is the water cycle, told as an allegorical tale. Warmhearted Eva with the warm hands evaporates the water. The water vapour is invisible so the Water God does not see the water leave the seas. Con jumps high into the air and condenses the water vapour with his cold breath and makes the clouds. Thus the processes of evaporation and condensation are the key elements of the story.

The Children of the Water God

Once ... in the very beginning, all the water in the world was in a single ocean. The Water God lived there in his ocean, keeping himself to himself.

Then humans were created. They lived on the land but they needed water to survive. Every evening, the humans would come to the edge of the water and ask the Water God for a little water and he would let them fill their buckets and pots so that they had enough to drink and wash and to water their crops.

Time passed and the number of humans grew and grew. They needed more and more water. This made the Water God angry. 'This is all mine!' said the Water God, 'And you are greedy! You cannot keep having more.'

The men came and collected their water every evening as usual, but now it was not enough so, at night, men crept down to the ocean to steal more.

Soon, the Water God noticed, and roared with fury, 'Who has been stealing my water? Enough! From now on you get no water at all.'

The Water God had two children, Eva, a warm-hearted girl with warm hands, and her brother, cool Con. 'You two will guard my water day and night,' the Water God shouted. 'If you catch any thieves, bring them to me!'

That night, brother and sister walked around the lake watching for thieves. It wasn't long before they caught the first man trying to fill his bucket. 'Please,' said the man. 'Let me take water or my family will die. Just a bucketful. There is plenty left here!'

'Sorry,' said Con. 'Our father will see soon enough, and then we will all be in trouble.'

'But then my children will die of thirst!' the man pleaded. 'Without water in our fields there can be no food and we will starve. Please, help us!'

Con's warm-hearted sister felt sorry for the man and she began to think. 'Maybe we can help,' she said. 'I have an idea. We could make the water invisible and then Father won't notice it going. It just might work!' She whispered the instructions to her brother and he listened and he nodded.

'Good!' he said. 'That's a cool idea!'

Eva picked up a giant handful of water in her warm hands and blew on it with her warm breath. As they watched, the water bubbled, evaporated and disappeared up into the sky as invisible water vapour.

Then cool brother Con jumped high in the air and breathed cool air on the water vapour and watched as it turned into tiny droplets of water hanging in the air. It was a cloud. Then he blew the cloud out over the land. As it was pushed higher into the colder air over the mountains, the cloud formed bigger droplets and they began to fall onto the ground as rain.

Every night, brother and sister played the same game and soon there were rivers and streams and enough water for all the humans. The Water God was suspicious. He came to watch his children. 'What are you doing?' he said as Eva blew on the water in her hands and Con made clouds.

'Oh we're just playing, Daddy!' cried Eva and Con. 'It's such fun! Do you want to play too?'

'Are the humans still stealing water from the ocean?' their father asked.

'No, Daddy, they have not stolen a drop from the ocean! We promise!'

The Water God was not fooled for long. He saw a sky full of clouds and the rivers flowing. He understood what his children had done. 'How dare you disobey me?' he roared. 'The water is mine! It all belongs to ME!'

'But, Father,' said Con. 'It still belongs to you. It's just doing more good now. Look, it turns into clouds and then to rain and then rivers and lakes, but in the end it all comes back to your ocean. You don't lose any of it. It just goes round and round.

'Listen, Father,' said Eva, 'we've even made a song about it ...'

We blow the water high
As vapour in the sky
When it cools it turns to rain
And falls to Earth again
Then streams and rivers flow
Down to the sea below
Never lost and always found
It just goes round and round and round

The Water God's heart softened when he saw how much his children cared about him and his water. He smiled and joined hands with his children. They danced and sang all through the night. So it has been and so it always will be.

P
PHOTOCOPIABLE FOR USE IN SCHOOL

Top tips for telling

This is a story about gods, so the first thing is to find larger-than-life voices for them. Think BIG! The Water God has the quality of hoarding and wanting to keep HIS water. His children are more playful and caring. Make sure the suffering of the humans without water is clear. The thirsty and hungry families are dying from lack of water.

The second dramatic moment is when the father finds out what his children have done. Stretch out the suspense as he is angry at first but then softens when he realises that he has lost nothing.

Do try to include the song. It adds fun and variety to the story and helps your class to remember the water cycle sequence. Any tune will do, or else you can chant it and then your students will find their own tunes when they retell the story. You could add actions to the song: hands flutter up, then flutter down, then a flowing movement and finally a gesture for the ocean.

Ways to work with the story

Teach the story using HMSS

Deepen the story with activities such as these:

- Re-enact the water cycle with the children as the water molecules. Stay close together when the water is a liquid and spread out when it is a gas.
- Role-play Eva and Con explaining to their father that the water isn't lost, it just goes around in a cycle. Explain evaporation and condensation and that the water vapour is invisible.
- Role-play being a TV weather expert explaining why it is snowing in the mountains.
- Explore the dilemma Eva faces when she has to choose whether to help the people by giving them water or to obey her father's instructions. You could use a thought corridor (see p. 66).
- Re-enact a family who are hungry and thirsty talking about why they have become so and what they should do.
- Re-enact a father of a thirsty family coming to plead with the Water God for water.

Explore the science that is directly linked to the story:

- evaporation and condensation are reversible changes;
- the water cycle;
- water is needed for human, animal and plant life.

Explore the science that is indirectly linked to the story:

- Dissolving and mixing materials can be reversible changes.
- Some changes are irreversible, e.g. mixing bicarbonate of soda and vinegar.

Try these science activities:

THE WATER CYCLE

1. Evaporation and condensation

Try putting a drop of water on the back of everybody's hand. Watch it. Where does it go? Ask questions such as 'Where does the water from the puddle in the playground go?' Introduce the idea that water can be an invisible gas. Breathe on mirrors. Discuss where the wetness comes from. Leave ice cold drinks out in the classroom – it gets wet on the outside of the glass. Discuss why this happens. Ask questions such as 'Why do we breathe on our spectacles to clean them?' and 'Why does the inside of the car windscreen steam up in winter?'

2. The water cycle

Discover that water evaporates and condenses to change state. You can put hot water in a pot with a lid of cling film with ice on the top – and see the water condense and drip back to make a water cycle. Draw the water cycle. Note the three states of water (solid ice, liquid water and gaseous vapour).

3. States of matter

Look at how solids, liquids and gases behave. Compare water pouring with sugar granules pouring. Look at both under a hand-held magnifier/microscope to see that the sugar is made of tiny pieces of a solid whereas water is a liquid. Note that liquids spread out to fill the bottom of a container whereas a spoonful of sugar can be heaped up. Play with gloop (cornflour and water) to see that it sometimes behaves like a liquid (you can pour it) and sometimes like a solid (you can snap it). Spray perfume in one corner of the room. It will eventually spread throughout the room, carried by the air, which is constantly moving. Relate all this to particle theory: the particles in solids are held together by strong bonds, in liquids the particles are able to flow around each other and in gases the particles are not bonded to one another at all.

4. Reversible changes – changes in state

With adult supervision, melt ice, butter and chocolate in a microwave or over a hob. Refreeze them to see that changes in state are reversible. It's fun to ask the children to devise a test to see which ice lolly would be best on a hot day by measuring how fast the ice lollies melt. The easiest way is to measure the length of the lolly as it melts.

Continued overleaf

THE WATER CYCLE *continued*

5. Reversible changes – dissolving salt

Dissolve salt in water. This is a reversible change too. You can leave salty water on a windowsill and observe the salt crystals form when the water evaporates. The process can be done quickly by heating a cupful of very salty water on a baking tray in an oven or in a saucepan on the stove. It doesn't take long, so make sure an adult is watching. Investigate what makes salt dissolve faster. Race hot against cold water, stirring quickly against stirring slowly, and crushed salt against whole salt flakes to change the particle size. Time how long it takes the salt to dissolve in each case to see that hot water, fast stirring and small particle size all speed up the process of dissolving. Discuss how to make this test fair.

6. Irreversible changes

Observe some permanent changes when materials are heated, mixed or burned.
For example you might toast/microwave marshmallows, burn indoor sparklers or make carbon dioxide using bicarbonate of soda and vinegar in a tall jar. The carbon dioxide will sink to the bottom of the jar and you can pour it out over a lit candle and show that it puts the candle out. This proves that a gas (not air) has been produced which escapes so the change is irreversible.

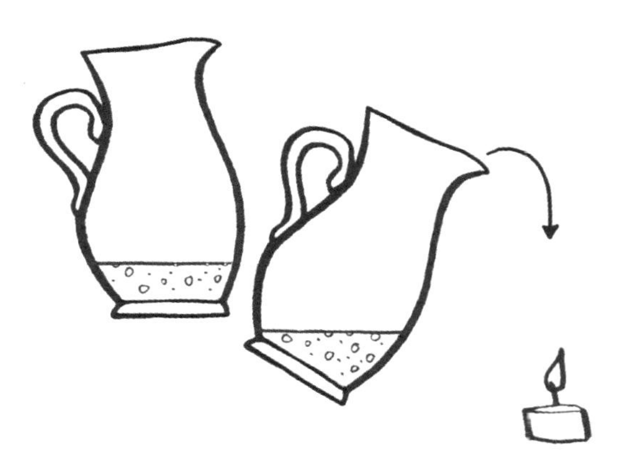

Explore other curriculum areas that link to the story:

Reading

Read other creation myths such as *Tiddalick, the Greedy Frog: An Aboriginal Dreamtime Story* by Nicholas Wu Pap (Teacher Created Materials, Huntington Beach, CA, 2013).

Read non-fiction texts about the water cycle.

Fiction

Box up and write the story, or innovate first, for example telling the story from the point of view of a human family, or from the point of view of the Water God. You could also innovate using crops or soil or worms as the main characters, begging the children for help.

Non-fiction

Instructions: Be Eva or Con and give instructions about how to make rain.

Explanation: Explain how the water cycle works. This could be done as a presentation or a poster.

Explanation: Be a biologist and explain why plants or animals need water.

Information report: Write a report on the features of a river or ocean.

Persuasion: Write a letter to the Water God persuading him to share the water, explaining why it's needed for life and how he loses nothing.

Poetry

Read and write poems about the sea, rivers clouds or rain. You could write these as poems containing kennings like 'wave-roller', 'spray-thrower', 'ship-rocker'.

Religious Education

Find out about the Native American practice of rain dancing.

Read the Bible story *Noah's Ark*.

Geography

Research rivers and oceans around the world. Learn about the stages of a river from its source to the ocean.

Make a rain gauge (out of a 2 litre drinks bottle) and measure the amount of rain collected each day. Create a class weather chart.

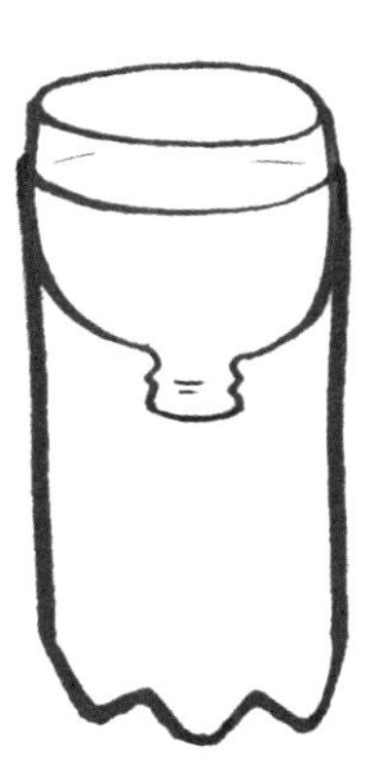

Research areas of the world that are suffering from drought and any ways that this problem has been tackled.

Learn about the causes of flooding (this may link to global warming). Research areas of the world that are suffering from flooding and the effects of this. Learn about flood defence and protective measures.

Find out about local flooding issues.

Maths

Measure temperatures using the Celsius scale which is linked to the freezing point and boiling point of water.

Use the data from your rain gauge to make bar charts of rainfall.

Compare your data to rainfall data in other countries.

Topic 4.6 MIXING AND SEPARATING MATERIALS

Story: Nimblefingers

Topic: Dissolving, heating materials, sieving and filtering
Plot type: Overcoming the monster
Genre: Wondertale

This is a story about a clever young girl who gets such a reputation for being able to fix things that people think she uses magic. Stolen away by a polar bear who refuses to free her unless she grants him three wishes, she finds a way to make his wishes come true using her scientific know-how.

The science in the story is revealed in the way Nimblefingers solves the problems. She uses evaporation to separate salt from seawater, filters the bathwater to make it clean and sieves dirt to find gold to make the bear his crown.

Nimblefingers

Once ... there was a girl called Blossom. She had long golden hair and the daintiest fingers you have ever seen. She could stitch and sew, she could spin and weave but her favourite thing of all was to fix broken things. She was very clever and she worked so fast you could barely see her fingers move.

Her parents were very proud and so if anyone told them that something was broken they'd call for Blossom and her nimble fingers to come and fix it right away. Blossom never said no. She enjoyed each new challenge and loved to make people happy.

Word spread. The people from the neighbouring village would stop by on their way to market to give Blossom something to fix as they went past.

'Blossom ... Could you fix my clock?'

'Blossom ... my spinning wheel is jammed!'

'Blossom ... with your nimble fingers could you get this key out of the lock?'

Pretty soon, she was so well known in those parts that people just referred to her as 'Nimblefingers'. 'Nimblefingers will fix it! Nimblefingers is magic!' they said. And the name stuck.

One dark evening, there was a loud knock at the door.

BANG ... the door flew open and in came a rather shabby and angry-looking polar bear. Blossom and her parents backed away from the huge creature.

'I ...WANT ... NIMBLEFINGERS!' he growled, baring his teeth.

Blossom stepped forward. 'I am she,' she said confidently. 'What do you want with me?'

'I WANT MY THREE WISHES FROM THE FAIRY NIMBLEFINGERS!' he roared and he picked Blossom up with a rough paw and threw her upon his back.

'She's not a fairy .... She's just good with her hands!' shrieked Blossom's mother.

'Blossom ... whatever can we do?' wailed her father.

'Don't worry,' called Blossom as she and the bear disappeared out of the door. 'I'll think of something.' And with that they were gone.

The bear travelled all night through the darkness over high mountains and through scratchy forests until at last he came to the sea. He pitched himself straight into the black water and Blossom clung onto his fur while he swam. It was dawn when they reached the island.

'Nimblefingers,' he said, 'I have heard that you know magic. I am king of this island ... but look at me. Nothing about me, or my palace, looks royal any more. Grant me three wishes, then I will take you home. Fail and I shall eat you for my supper.' Then he growled, baring his teeth.

'What do you wish for?' asked Blossom, hoping desperately that she'd be able to meet each challenge.

'First, I wish for salt for my food. I have no money to buy any and my food tastes terrible. I wish for a jar of salt that never runs out!

'Secondly, I wish for clean, white fur not muddy, matted fur. I want to be clean and white like a king should be.

'And, thirdly, I want a crown, a golden crown. Give me a golden crown so I shall look like a king.'

Blossom nodded and went away to think. The first wish was easy. They were on an island by the sea. She scooped up some seawater in a cooking pot and lit a fire beneath it. Pretty soon, the water had all evaporated away and she could scrape the salt crystals from the bottom of the pot. She placed them in a jar and gave them to the bear.

He seemed pleased. He was a little less growly. Every night, she took her cooking pot and boiled up some seawater until all the water had evaporated away and then she put the salt into the bear's jar and in this way she kept it topped up. Every morning the bear found it full. He thought that Nimblefingers was magic, so he thought the pot was magic.

'What about my fur?' he grumbled.

Blossom had noticed that the bear bathed every night in a bath of dirty river water, which left him muddy and mucky.

'I shall need a fine silk thread and loom,' she said.

Blossom wove the silk thread into a delicate cloth. She took the cloth to the tap that filled the bear's bath and covered the spout with a few layers of cloth. When she turned on the tap, the grimy, brown river water ran through the cloth and came out clean.

Lying down in the clean water, the bear wriggled and writhed with pleasure. The dirt from his fur loosened and he was left snowy white, as a royal bear should look. Blossom changed the silk filter on the tap every night and took away the dirty one. The bear was pleased.

'What about my crown?' he asked with a smile.

The third wish was a little more difficult. Blossom simply didn't have any gold. But she had noticed that there was gold on the island. The palace had once had a golden roof made of golden tiles but it had been worn away by time and bad weather. The birds had pecked at it and now all the gold tiles had crumbled away, leaving a stone roof. But in the ground around the palace, Blossom had seen tiny shining shards from the old gold tiles. The gold was there. Blossom just had to get it with her nimble fingers. The great big paws of the bears were much too clumsy.

Blossom made a sieve out of wire with a wooden frame, making sure the holes were too small for the tiny shards to fall through. Next, she dug up shovelfuls of the soil and put it in the sieve, then shook it and shook it. Most of the soil fell through the holes but the gold shards and the big stones were left behind.

For days and days she shovelled and sieved until she had enough gold for the crown. Then she washed the gold and heated it up on the fire in a pot until the gold melted to liquid. Straight away, she poured it into a mould. When it had cooled, she broke open the mould to reveal a small crown.

Blossom solemnly put the crown on the bear's head. He looked different now. His fur was white and his belly was full. He looked very regal in his new crown and ... maybe even happy.

Blossom was pleased. She took a deep breath and tried to look brave.

'I have granted your three wishes, your majesty. Now you must keep your promise and take me home,' she said.

The bear looked at her for a while. Then he sighed. 'Kings must keep their promises, I suppose.'

And with that he scooped her up and threw her onto his back. He swam back across the sea and went back through the scratchy forest, over the high mountains to Blossom's home.

Blossom was so happy to see her parents again.

'We knew you'd think of something!' they said as they hugged her tightly. 'You can fix anything!'

When the villagers heard what she had done they put up in the town square a statue of her riding on the back of a great polar bear and they sang songs about her for years to come.

Nimblefingers rode on a bear
Over the mountains and salty sea
She cleaned him up, made him a crown
And then she came back to you and me!

Top tips for telling

This is a story where the courage and cleverness of Blossom save the day. You need to find a voice for her which evokes her love of fixing things. She is confident and smart and willing to have a go at anything. The bear has to be scary and dangerous so we know she is in danger from such a strong and grumpy creature.

For the science elements (filtering, evaporating, sieving and melting the gold) try to describe the details of each process so the children can see it in their imagination.

At the end, show the happiness of the family and the town as they sing Blossom's praises.

Ways to work with the story

Teach the story using HMSS

Deepen the story with activities such as these:

- Make up scenarios in which people come to Nimblefingers with a problem and she solves it. Then act it out. For example, the people might bring her a tangled marionette and she cuts off the strings and re-strings it so that it can be used again. The children can come up with problems to solve and the solutions and then act out the scenario.
- Re-enact the three problems solved by Nimblefingers with real equipment.
- Role-play Nimblefingers telling her parents how she managed to solve all the bear's problems and escape.
- Role-play Nimblefingers talking to the bear about his 'wishes' as she solves them one by one. Show how the bear changes from being grumpy to being content when he has his crown.

Explore the science that is directly linked to the story:

- mixtures can be separated by filtering, sieving and evaporating.

Explore the science that is indirectly linked to the story:

- Evaporation and condensation are reversible changes.
- Dissolving, mixing and changes of state are reversible changes.
- Some solids will dissolve in liquids.

Try these science activities:

MIXING AND SEPARATING MATERIALS

1. Dissolving

Test a variety of different materials to see if they will dissolve in water: salt, sugar, flour, sand, paint. Discuss what happens and introduce the idea of dissolving.

2. Mixing materials

Try mixing butter and water or oil and water. Note how some materials will not mix however hard you try. Mix materials that do something unexpected, such as bicarbonate of soda and vinegar or plaster of Paris and water. Discuss whether these materials can be separated again.

3. Separating salt from saltwater

Demonstrate that salt can be separated from saltwater. Talk about days at the seaside and how we can feel the salt on our faces on the way home. Leave saltwater on the windowsill in a shallow dish. Watch the salt crystals form as the water evaporates away. You can also do this by heating a dish of saltwater on the hob and watching the salt crystals appear before your eyes. It can also be done by heating a baking tray of salt water in the oven. Link this to Blossom's first task.

4. Separating materials – sieves

Separate materials using sieves. You could even ask the children to design their own sieve to separate particles of different sizes, e.g. pasta, rice and flour. A flat paper bag with the right-sized holes cut in the bottom is a possible design. Link this to Blossom's last task.

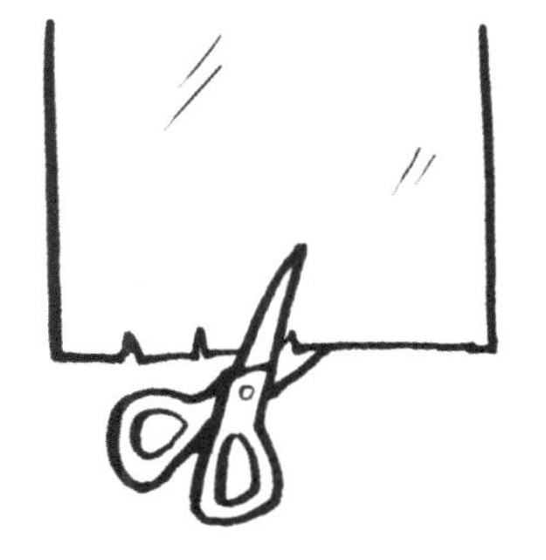

5. Separating materials – filters

Once the children are familiar with the idea that some solids will dissolve in water, introduce the idea of filters – show them coffee filters and teabags to encourage them to see that larger particles, which are not dissolved, are too big to pass through a filter, so tea leaves and coffee grains get filtered out. Give the children a range of mixtures to separate, e.g. pasta and rice, salt and sand, sand and water. Let them work out how to separate them using the equipment you provide (colanders, sieves and filters).

6. Separating materials – filter challenge

Ask teams of children to make their own filter in a funnel using materials they have chosen. (Top tip: cotton wool is pretty good if pressed quite densely into the funnel.) Mix a few spoonfuls of soil into a big jug of water to make some dirty water. Once the teams have made their filter, put them to the test! Pour 100 ml of the dirty water through the filters and see if the water comes through clean. The cleanest water filtered from a jug of muddy water is the winner! We like to line up the pots of filtered water in order of cleanliness and then discuss which materials were used in the corresponding filter. It is usually clear that some materials make better filters than others and that the best filters have closely packed fibres. Link this to Blossom's second task.

7. Animals and their habitats

Find out about different bears and their habitats. Write a presentation about your chosen bear and present it to the class.

Explore other curriculum areas that link to the story:

Reading

Read other bear stories. Compare and contrast the different temperaments and roles played by the bears. There is a lovely version of *Goldilocks and the Three Bears* in *The Nursery Storybook* by Georgie Adams (Orion, London, 1996).

Read other traditional tales that involve wishes. Compare and contrast them to *Nimblefingers*.

Fiction

Box up and write the story, or innovate using new problems involving different mixtures for Nimblefingers to separate.

Non-fiction

Instructions: Write instructions on how to make a sieve, silk filter or golden crown.

Information report: Write reports on bears, sieves and filters, or panning for gold

Persuasion: Advertise your filter – this could be an advert for TV.

Persuasion: Write an advert for Nimblefingers' Fixit Shop.

Persuasion: Write a letter from Blossom's parents persuading the bear to let her go.

Explanation: Explain why you need different sieves for different purposes, e.g. colander, sieve, tea-strainer or cafetière plunger, in a written explanation with diagrams.

Design/Technology

Cooking is a great way to show various changes. Talk about the changes that are happening as the children cook something. You could make chocolate crispy cakes with them to show how butter and chocolate can be melted and then cooled back to a solid. Make sure that children are closely supervised by an adult when melting chocolate.

Chapter 5

Physics Stories

Topic 5.1 FORCES – PULLING AND PUSHING

Story: The Giant Turnip

Topic: Forces – pulling and pushing
Plot type: Quest
Genre: Folktale

This is a traditional story about a farmer who is unable to pull up a turnip on his own so he asks everybody to help him to pull. Only when the whole household pull together, combining their force, does the turnip finally come up.

The science content in the story introduces the idea of using a force, in this case a pulling force, to move things.

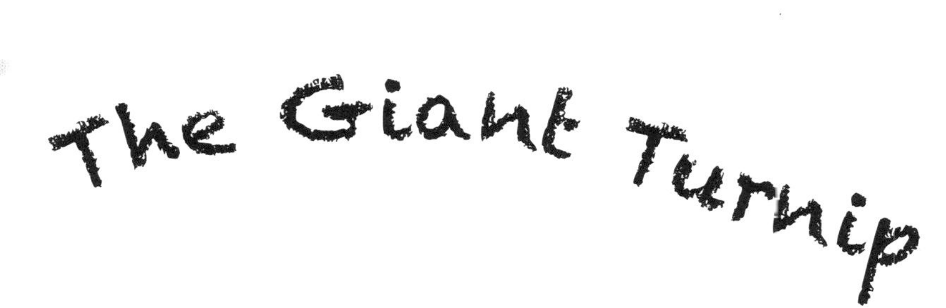

The Giant Turnip

Once upon a time there was a farmer. He lived in a house with his wife, his son Jack, his sheepdog, and a mouse.

(*Include any other family members or animals you like at this point.*)

One day he took a seed and planted it in the ground. Every day he watered it.

The roots went down and the shoots went up.

The seed grew and grew and grew until it was the size of a man's head and still it kept growing.

One day, the farmer's wife said, 'Pull that turnip up! I want to cook it.'

So the farmer went out and took hold of the turnip top and pulled and pulled. As he pulled he sang this song:

Heave-ho, heave-ho! Pull the turnip free.
Heave-ho, heave-ho! We'll have it for our tea.

He sang and he pulled, he pushed and twisted, but the turnip just would not budge.

He called out to his wife. 'Can you help me to pull?'

And the farmer grabbed the turnip, the wife grabbed the farmer and together they pulled and sang:

Heave-ho, heave-ho! Pull the turnip free.
Heave-ho, heave-ho! We'll have it for our tea.

They sang and they pulled, they pulled and they sang. But the turnip would not budge.

The wife called to her son, Jack, 'Come and help us pull.'

The farmer grabbed the turnip, the wife grabbed the farmer and Jack grabbed his mother and together they all pulled and sang:

Heave-ho, heave-ho! Pull the turnip free.
Heave-ho, heave-ho! We'll have it for our tea.

They sang and they pulled and they pulled and they sang. But the turnip would not budge.

(Repeat for the dog and the mouse and anything else you like until you have everybody in the story helping to pull.)

The farmer grabbed the turnip, the wife grabbed the farmer, Jack grabbed the wife, the sheep dog grabbed Jack, the mouse grabbed the dog and they all pulled and sang:

> *Heave-ho, heave-ho! Pull the turnip free.*
> *Heave-ho, heave-ho! We'll have it for our tea.*

– and this time the enormous turnip popped out.

The wife chopped it up, put it in a stew and that night they all enjoyed a delicious bowl of turnip stew.

Afterwards they sang:

> *Heave-ho, heave-ho! We pulled the turnip free.*
> *Heave-ho, heave-ho! We've had it for our tea.*
> *Heave-ho, heave-ho! Now it's time to go.*

Top tips for telling

This is a simple repeating story about working together and pulling together. The main character is the farmer, who is determined to pull the turnip up for his wife. You need to show this determination for the story to work. This can be demonstrated in the physicality of pulling as hard as one possibly can. Get the class to join in with the pulling and the singing or chanting. If you like, take volunteers to play the turnip, farmer, wife and so on so that the class sees embodied how the story develops. Make sure you celebrate the turnip popping out and savour the lovely taste of the soup.

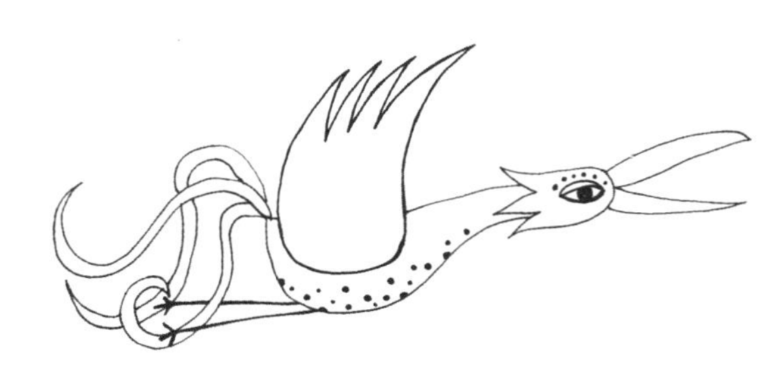

Ways to work with the story

Teach the story using HMSS

Deepen the story with activities such as these:

- Role-play the farmer explaining to his wife that he can't pull hard enough to move the turnip himself and needs her to pull too, so increasing the force.
- Explore the movement of pulling in the story in mime or dance.
- Make a gallery of pictures of the different stages in the story.
- Role-play a TV news reporter reporting from the scene.
- Role-play the mouse telling his family how his little extra pull made the difference.

Explore the science that is directly linked to the story:

- Push, pull and twist are all forces.
- Forces can make things begin to move (if they are not already moving), speed up, slow down, stop or change shape.

Explore the science that is indirectly linked to the story:

- Gravity is a force that pulls you down towards the Earth.
- Blowing is a push force.
- Two magnets will exert a force on each another.

Try these science activities:

FORCES – PULLING AND PUSHING

1.Tug of war

Play tug of war so that you can feel the effect of someone pulling you. Play in pairs first, one against one in a big space on grass or a soft surface. Then add more people to one team so that the contest becomes unbalanced. See if they can make a stronger pulling force when they all pull together. Ask questions like 'Does it make a difference if you wear school shoes or trainers?' and test them out as long as what that involves is safe! Draw pictures of what happened. Take care to leave spaces between children in each team in case someone slips.

2. Moving cars

Play with free-rolling toys. Push them along and pull them along. Try to move them in different ways, e.g. roll them down a slope or blow them along. Draw the ways that you moved the car. Note that pushes, pulls and gravity can all make a stationary car move.

3. Magnetic toys

Play with magnetic toys to feel them pulling each other along and pushing each other away. You can even use toy trains with magnetic connectors to show how the magnets will make the train move by attracting (pulling) or repelling (pushing). Note that, just like pushes and pulls, magnetic forces can make things move too.

4. Blow football

Play blow football with two straws and a ping-pong ball. Put the ball on a table with two goals marked at opposite ends. Try to blow the ball through your opponent's goal before they blow it into yours. Note that blowing and wind are push forces too. You could make boats with different-sized sails and try blowing them along.

5. Making kites

Make simple small kites and pull them along in the playground until they catch the wind. You can try making kites without tails too so that you can see how the tail keeps the kite correctly oriented by pulling one end down. Draw the forces at work on your kite.

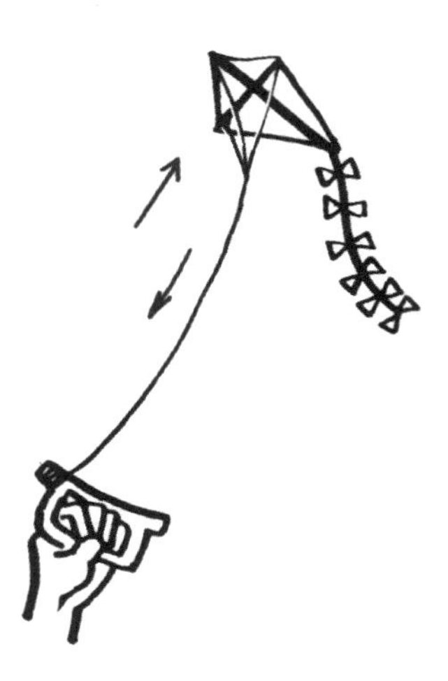

6. Shaping dough

Make shapes with clay or dough. Pull the dough to stretch it out and push the dough in to leave dents. Note that the forces haven't made anything move but they have changed the shape of the dough. Take a photo of your dough shape. Annotate it with arrows to show where you pushed it and where you pulled it.

Explore other curriculum areas that link to the story:

Reading

Read other stories about pushing or pulling, e.g. *The Elephant's Child* in *Just So Stories* by Rudyard Kipling (Oxford University Press, Oxford, 2009) or *Brother Whale and Brother Elephant* in *More Ready to Tell Tales*, edited by David Holt and Bill Mooney (August House, Little Rock, 2000).

Fiction

Box up and write the story, or innovate pulling up a different vegetable or introducing different family members and pets to help with the pulling.

Non-fiction

Instructions: Write instructions on how to grow/pull up a turnip.

Instructions: Write a recipe for turnip soup.

Recount: Write a letter to Grandma about how you helped to pull up the turnip.

Persuasion: Write an advert for a turnip-pulling-up machine. If you are working with younger children, you could invent the details for the machine as a class and then write individual adverts.

Persuasion: Be the farmer and persuade reluctant family members to help pull up the turnip.

Explanation: Be the farmer and explain why you needed everyone to help.

Art

Make clay models of the people in the story. As the children work the clay, talk about the pushes and pulls they are using to shape the clay.

Design/Technology

Make a vegetable stew or soup from root vegetables that have to be pulled up like a turnip. If you have a school vegetable garden you could pull up your own vegetables for the soup.

Topic 5.2 LIGHT AND NIGHT AND DAY

Story: The East and the West

Topic: The orbit and rotation of the Earth, day and night, light and how we see

Plot type: Quest

Genre: Creation myth

This is a myth set at the beginning of time. The Earth has to be turned by the Sun God in order for the people on both sides to have some sunshine. Fed up with all this hard work, the Sun God sets the world spinning around the sun, causing day and night, and this solves the problem.

The science content in the story includes the way that the Earth's rotation causes day and night and that the sun is in the centre of our solar system and the Earth moves around it. It also shows that we need light in order to see and for plants to grow.

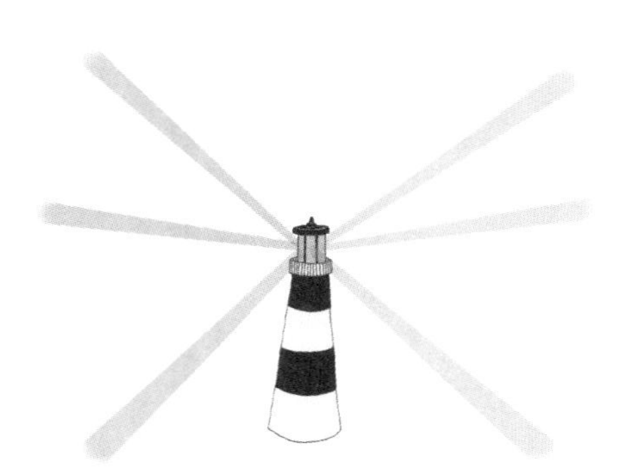

The East and the West

Once ... at the beginning of time, planet Earth was still. The gods had carefully placed it into the universe and there it sat in space, not moving at all.

The sun was in the East. It shone all the time on the Eastern side of the Earth, while the West was always in the dark. The people of the West complained to the Sun God:

'We need the sun for our crops to grow.'

'It's too cold in the West – we need the sun to keep us warm.'

'It's too dark – we can't see where we are going.'

As they stamped their feet to keep warm they sang:

Sun God, Sun God
Bring us the sun
It's dark and cold, we need your light
Sun God, Sun God
Bring us the sun
Nothing can grow, please make it right

The Sun God was trying to sleep when he heard the people stamping and singing. 'I wish they would be quiet,' he said, 'so I could get some sleep.'

He stretched out his golden arm and gave the Earth a quick twist so that the world faced the other way.

'Now maybe I can get some sleep,' he said and he closed his eyes.

Now the West was in the sunshine and the Westerners were happy. 'Ooooh lovely! Thank you, Sun God. Now we can see and grow and stay warm!'

But the Easterners were furious! They hated being in the dark.

'Hey!' they shouted. 'It's cold. We can't see where we are going and our crops will die.' They shook their heads, stamped their feet and began to sing:

Sun God, Sun God
Bring us the sun
It's dark and cold, we need your light
Sun God, Sun God
Bring us the sun
Nothing can grow, please make it right

The Sun God opened his eyes again. 'I'll never get any sleep,' he said, scowling. He reached out his golden arm and gave the Earth a quick twist, back to where it was before, and closed his eyes.

Now the East was in the sunshine again and the Easterners were happy. 'Thank you, Sun God,' they said.

But the Westerners were furious! They hated being in the dark. It was cold and they couldn't see where they were going and their crops stopped growing. They shook their heads and stamped their feet and sang

Sun God, Sun God
Bring us the sun
It's dark and cold, we need your light
Sun God, Sun God
Bring us the sun
Nothing can grow, please make it right

And so it went on. When the Earth was twisted one way the Easterners moaned.

When it was twisted the other the Westerners moaned.

The Sun God was fed up. The Earth was heavy and all the twisting made his arm ache. And he couldn't get any sleep at all with all that stamping and singing.

'Will you all be quiet?' he roared and he picked up the whole Earth and spun it away like a great big spinning top, off into space. It spun in a great big circle around the sun, spinning in space as it circled the sun.

When the Western side of the Earth spun round into the sunshine, the Westerners were delighted. Then, as the Earth spun round, they found themselves in the dark again. They were about to complain when someone said, 'Look we're spinning! If we keep spinning we will spin into the sunshine again soon.' They went to sleep through the dark hours and when they got up they had, indeed, spun into the sunshine again.

The Easterners on the other side had realised the same thing. They spent half the day in the dark and half the day in the light. The East had their fair share and so did the West. They were happy. And they stopped moaning.

As for the Sun God, he closed his eyes and fell into a long deep sleep.

And that is why, to this day, it is sometimes day and sometimes night because the Earth is still spinning in space.

Top tips for telling

For a creation myth like this you need to evoke the atmosphere of godlike beings, vast and powerful. They need big voices and huge emotions. If you have a drum you might play it first to get in the mood.

After that, the main thing is to be really clear about the directions of things. You might use a lamp and a tennis ball to model the positions of the Earth and the sun. Remember that the Earth is rotating and orbiting the sun at the same time, so you'll need to model this clearly. You could show this as you tell the story.

Make clear the people's anguish when left in the dark and why it is a matter of life and death. Then make clear their relief when the sun comes back.

Ways to work with the story

Teach the story using HMSS

Deepen the story with activities such as these:

- In a large space, practise moving like the sun and the Earth. The children being the sun need to stand in the centre of the room while the children being the Earth rotate and walk in a circle around them. You could even have someone be the Sun God and spin the Earth into motion.
- Re-enact the people complaining to the Sun God.
- Role-play a farmer explaining to the Sun God that they need light because the plants won't grow in the dark.
- Role-play the Sun God telling the other gods about his problem of the moaning people and his solution.

Explore the science that is directly linked to the story:

- Plants need water, light and a suitable temperature to grow and stay healthy.
- We have day and night because the Earth is rotating.
- We need light to be able to see.

Explore the science that is indirectly linked to the story:

- the structure of the eye and how we see;
- the sun is a source of light.

Try these science activities:

LIGHT AND NIGHT AND DAY

1. Planting seeds in light and dark places

Plant some bean seeds and watch them germinate. Put some, but not all, of the seedlings under a bucket or thick black plastic bag. Talk with the children about what they expect to see. Write their ideas in thought bubbles and keep these safe to see if they were right. Take care of both sets of plants. In a few weeks, the plants in the dark will have become paler and may be taller as they try to grow upwards to find sunlight.

2. Growing plants in different seasons

Sprinkle some cress seeds on tissue paper and water them. Put one set in the fridge and the other on a warm windowsill. Use suitable thermometers to measure the temperature in each place. Keep the tissue moist and take photos of the cress each day for a week or so. Relate this need for water and warmth to the seasons – nothing grows in the winter because plants need light and warmth to grow. Draw pictures or label photos of the cress growing or use a time-lapse photography app to film it.

3. Finding out how the sun helps our food to grow

Find out about all the foods we eat that are grown in the ground. You might like to tell the story of *The Little Red Hen*, since it tells how bread comes from a plant. Pie Corbett does a choral version which you can find on YouTube at www.youtube.com/watch?v=JdvJZD-cplg. You could visit a farm to see some crops or if you have a school vegetable patch you could harvest some vegetables. Find out how the food we eat is grown and why we need the sun for that to happen.

4. Making shadows

On a sunny day, spend time outside observing shadows. Encourage the children to see how their shadow begins at their feet. See if they can create a gap between themselves and their shadow by jumping. Stand them in a line so that they see that their shadows all fall to the same side.

Mark a spot on the playground and show them some objects. Tell them you will place the object on the spot, but first they must guess where the shadow will fall and what shape it will be. Let them mark their guess in chalk. Place the object on the spot and see if they were right. After a few goes they will start to make more accurate predictions.

5. Watch the world turn

Use a tree or post as a marker and track the shadows over the course of the day by outlining the shadow of a fixed object with chalk. Draw a big chalk compass on the playground. Mark where the sun is in the sky, every hour, by drawing an arrow on the chalk compass in the direction of the sun.

6. What can we see without light?

Make a dark space – use a blackout tent or create a dark den with blankets and cloths. Put torches and shiny things in the den. Allow the children to spend time in the dark and investigate what they can see. Help them to see that shiny things reflect light so you can only see them when the light is on, whereas light sources give out light.

Explore other curriculum areas that link to the story:

Reading

Read other creation myths about how things came to be, such as *The Crab Who Played with the Sea* from *Just So Stories* by Rudyard Kipling (Oxford University Press, Oxford, 2009).

Fiction

Box up and write the story, or innovate adding new reasons for the people to complain, e.g. the children are afraid of the dark. You could also change the story around and have people complaining they get too hot and can't sleep in the light.

Poetry/Music

Write new chants for the people to sing in your innovated versions of the story.

Non-fiction

Instructions: Write instructions on how to grow a healthy plant.

Recount: Write a letter from the Sun God or the people of the East recounting what happened to them.

Information report: Write a report entitled 'The Solar System'.

Information report: Write a poster about growing plants.

Explanation: Be the people of the East and explain why you need the light.

Discussion: Discuss the question 'Should the Sun God let half the people die?'

Maths

This is a good opportunity to talk about sharing. You could link this to division. Try reading your class *Give Me Half!* by Stuart J. Murphy (HarperCollins, London, 1996), since this links sharing food with the concept of halving and dividing.

Geography

Use compass directions in simple mapping tasks such as moving around the school grounds following directions on a map, e.g. 'At the back door, turn to the west then go four steps forward.'

Look on a globe or a map to find out which countries are to the east of where you live, and which are to the west.

Find images or live streams from satellites on the internet which show the Earth as seen from space, with some countries in darkness and others in light.

Computing

Use programmable floor robots. Give them directions to move them around a course. Draw a big map with the east and the west marked on it so that the children become familiar with the compass directions.

Religious Education

This story resolves with both the people of the East and the people of the West having their fair share of the sunlight. Discuss the things that your class have to share, such as the playground toys or colouring pencils. Talk about how it feels when someone won't share with you.

Read Bible stories such as the story Jesus told about *The Widow's Mite*. The widow gives a tiny amount to the temple but it is all she has to give. Jesus explains that because this is a huge amount for her to give away it is a bigger gift than the large sum of money from the rich man. Discuss how sharing what you have can affect the lives of others in a positive way.

Topic 5.3 SOUND AND HEARING

Story: Little Rabbit Goes Home

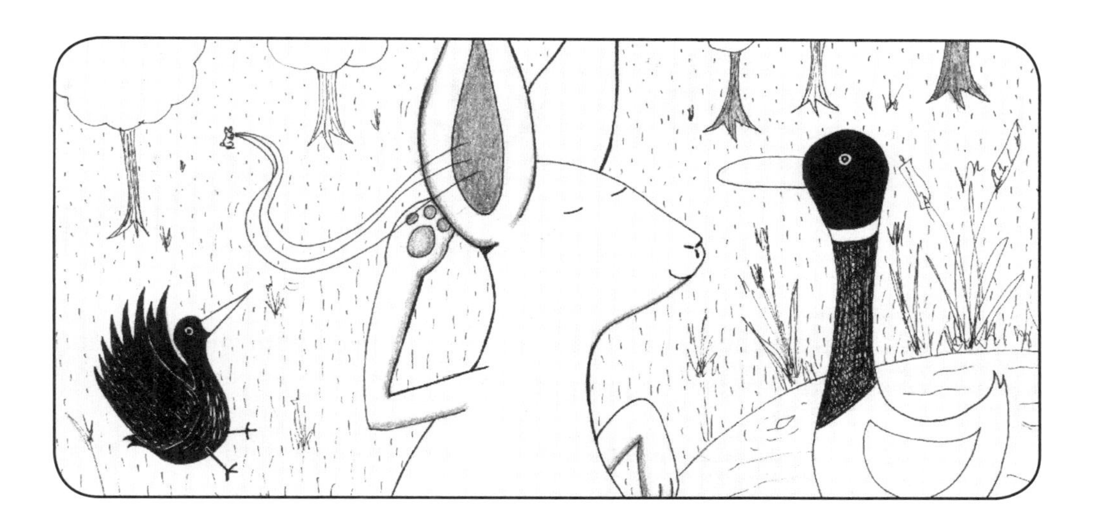

Topic: Sound
Plot type: Voyage and return
Genre: Fable

This is a story about how a little rabbit gets lost, meets a duck, a bird and a snail who give him poor advice, before an owl helps him learn to use his own abilities.

The science content in the story is all about sound, hearing and the fact that the closer we get to the source of a sound, the louder the sound. It could also be used to illustrate that different animals live in different habitats. The story is also a fable about self-esteem and knowing one's own abilities.

This story could also be used to support the teaching of habitats in conjunction with the planning suggestions given for Topic 3.4.

Little Rabbit Goes Home

Once upon a time there was a little rabbit. He lived with his mum and dad in a burrow under the ground.

One day he went for a walk in the forest and got lost. Little Rabbit was frightened and started to cry. 'I want to go home!'

Along came a bird.

'Why are you crying?' said the bird.

'I'm crying because I'm lost and I want to go home,' Little Rabbit replied.

'Oh you poor thing! When I'm lost, I flap my wings and fly high in the sky until I can see my home, then I fly there.'

'Thanks,' said Little Rabbit, 'I'll try that.'

So Little Rabbit jumped up and down trying to flap his front legs as if they were wings.

FLIP FLAP FLIP FLAP FLIP

He tried and tried but he couldn't fly. Little Rabbit started crying again.

'I'll never get home like that!' he cried.

Along came a duck.

'Why are you crying?' said the duck.

'I'm crying because I'm lost and I want to go home.' said Little Rabbit.

'Oh you poor thing!' said the duck. 'When I'm lost I jump in the river and swim until I can see my home.'

'Thanks,' said Little Rabbit, 'I'll try that.'

So, Little Rabbit jumped in the river and kicked his little legs trying to swim.

SPLISH SPLASH SPLISH SPLASH SPLISH

But he couldn't swim and he sank down in the water. Little Rabbit tried and tried but it was no good, so he scrambled out of the river. He started crying again.

'I'll never get home like that!'

Along came a snail.

'Why are you crying?' said the snail.

'I'm crying because I'm lost and I want to go home.'

'Oh you poor thing. When I'm lost I just make myself small and then go into my home on my back, then I'm home.'

'Thanks,' said Little Rabbit, 'I'll try that.'

So Little Rabbit curled himself up into a tiny ball and squeezed himself as small as he could.

SQUEEZE, SQUISH, SQUEEZE, SQUISH, SQUEEZE

But it didn't work. He didn't find his home. He started crying again.

'I'll never get home like that!'

Along came an owl.

'Why are you crying?' said the owl.

'I'm crying because I'm lost and I want to go home. I tried flying like a bird, swimming like a duck and shrinking like a snail, but it didn't work. I want to go home!'

The owl smiled. 'You aren't a fish or a duck or a snail. You are a rabbit with great big ears. You've got the best ears in the forest. Use them!'

Little Rabbit sat on a log, closed his eyes and listened very carefully. He heard the sound of the birds singing, the ducks quacking, the snails slithering, leaves rustling, wind whistling ... all the sounds of the forest.

Little Rabbit heard his mum's voice in the distance: 'Little Rabbit, Little Rabbit, where are you?'

He ran towards the sound. It got louder and louder and then ... there was his mum. They went down into their burrow. And had a big hug.

Top tips for telling

For the story to work you need to evoke the fear and desperation of the lost bunny, and then the hope each time he meets an animal who gives him some advice. There are three cycles of disappointment and then the magic moment when he hears his mum's voice and knows his own strength. In between, you can get the class to join in with the movements and sounds as he tries to fly, swim and squish himself up into a ball. Make sure you feel the joy of the hug at the end!

For early years the story could easily be retold with puppets or objects. Make sure everyone knows the animals before you start.

Topic 5.3 SOUND AND HEARING

Alternative story: The Blind Man and the Hunter

Topic: Sound and the sense of hearing
Plot type: Rebirth
Genre: Folktale

This is a traditional tale about a wise blind man and a proud hunter. The hunter sees no worth in the blind man because he cannot see the animals or shoot an arrow to hunt. He finds out that the blind man has learned to listen to the world so carefully that he is able to warn the hunter of dangers. The hunter tries to trick the blind man but the blind man knows he has been deceived. In the end, the hunter changes his ways.

The science content in the story includes the fact that we have senses that tell us about the world. In this story, the blind man relies heavily on his sense of hearing and smell, since he does not have the sense of sight. The tale highlights how useful these senses can be.

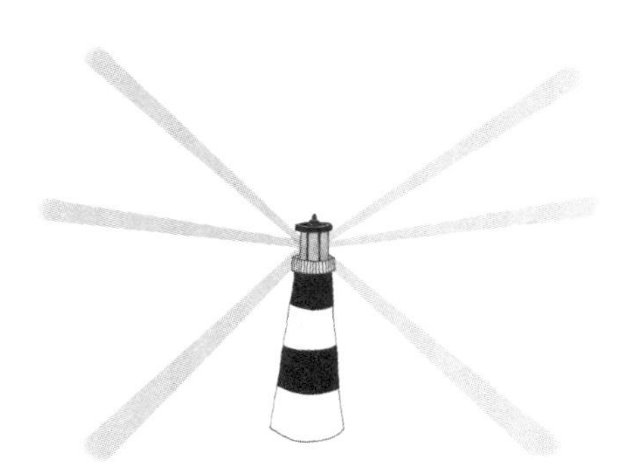

The Blind Man and the Hunter

Once ... there was a forest and in the forest was a village and in that village there were huts woven from branches and earth. In one of these huts lived a blind man. The man was greatly respected in the village. It was said that although he could not see with his eyes he could see with his ears. People came to him with their problems and he listened and gave them wise advice. He shared his hut with his sister, who looked after their home and looked after a small piece of land where they grew vegetables and kept a few goats for milk.

Then the sister got married and her husband, a hunter, moved in.
Every day, the hunter would go out to the forest, trapping and tracking, and every evening he'd come back with his kill and hand it over to his wife for cooking.

One day, the family was eating supper. The hunter was chatting away about his adventures in the forest when the blind brother interrupted, 'Let me go hunting with you tomorrow. I can help you.'

The hunter laughed. 'What, you? What possible use could a blind man be to me?'

The blind man shrugged and went back to his supper.

The next day, the blind man asked again, 'Let me go hunting with you tomorrow. I can help you.'

'Look, I've told you. You can't help. I am fine. I go hunting. I bring the food. She cooks it and you eat it. That's it. You can't come.'

Every day for one hundred days the brother asked to go hunting, but he was always dismissed as being of no use until finally the hunter relented. 'Okay, brother, tomorrow you can come. Just once. Then maybe you'll stop bothering me.'

The next day, the two men set off into the forest. For a while the hunter sang his hunting song as they walked, and the blind man followed him easily along the path without help. After a while, the blind man said, 'Shhh! Careful! There's a lion with her cubs nearby, over there that way.'

The hunter stopped singing. 'How do you know?' he asked doubtfully.

'I can hear them playing.'

The hunter listened but could only hear the wind in the trees. He peeped around the corner and, sure enough, lying there in the sun was a lioness with a litter of suckling cubs. Very dangerous! The two men tiptoed away.

After a while, the blind man said, 'Shhh! Careful! There's a mamba snake on the branch above your head. Move away slowly or he may strike.'

The hunter looked up and saw the snake poised ready to strike. He moved away quickly.

'How did you know?' he asked, impressed. The hunter was beginning to believe the stories he had heard about the wisdom of the blind man. He did seem to know things about the world.

'I smelt its scent,' answered the blind man.

The hunter went back to his song, but soon the brother stopped for a third time. 'There's a family of elephants down by the river. Should we go the other way?'

The hunter listened but couldn't hear the river, let alone the elephants. They walked for a good ten minutes before coming to the river, where a family of elephants were playing and washing. Now the hunter was really impressed. 'Maybe a blind man can be of use after all,' he thought, but did not say it.

They came to a clearing and the hunter showed the blind man how to set a bird trap, guiding his hands as he looped the thread around the taut bent stick. The hunter set one trap and the blind man another. Then they went on their way and went hunting with bow and arrow. The brother listened out for birds and animals to shoot and the hunter watched. The blind man was really useful, the hunter realised.

That evening at supper, the hunter said, 'Brother, come hunting with me tomorrow. Let's see what we have caught in our traps.'

The blind man nodded and smiled.

The next morning they came to the clearing where they had set the traps. There was a bird in each. In the hunter's trap was a plump brown bird and in the blind man's trap a bird about the same size but covered in the most beautiful feathers, all the colours of the rainbow.

The hunter liked the coloured bird more. He felt a little jealous. 'Well done!' he said to the brother as he took the birds out. 'You have caught a bird on your first try. Here, take it.'

He handed the blind man the brown bird. The blind man nodded and said nothing. The hunter took the rainbow bird as his kill and together they walked home. Somehow the hunter felt uneasy all the way home. As they were approaching the village, he stopped. 'Brother,' he said, 'You are said to be wise so tell me ... Why is there so much trouble in the world?'

The blind man replied quietly, 'It is because of people like you, who do what they should not!'

The hunter flushed with shame. 'I am sorry, brother. I was selfish. I thought you wouldn't know or wouldn't care, but it was wrong. Here – take the rainbow bird.'

The brother nodded and they exchanged birds. They walked for a while longer and the hunter stopped again. 'So tell me. So many people take what is not theirs. Is there any hope for the world?'

'Oh, yes!' said the blind man, smiling. 'There is hope. Because there are people like you who can admit their mistakes and learn from them.'

From that day on the two men hunted together and became the best of friends.

Top tips for telling

The characters of the hunter and the blind man are important here. The hunter is arrogant and the blind man is wise. Make sure the contrast is clear in your voice when they speak. Next comes the drama of the walk, when the blind man hears what the hunter cannot. Make sure you show the amazement of the hunter, and how he starts to respect the blind man. In the final scene, show how uncomfortable the hunter feels and how he repents of his selfishness.

Ways to work with the story

Teach the story using HMSS

Deepen the story with activities such as these:

- Role-play being the blind man walking through the forest. Imagine all the animal and forest sounds you would hear. Ask the children to describe all the different sounds to their talk partner.
- Role-play being the lost rabbit and imagine all the sounds you can hear. Describe them to your partner.
- Divide the story into parts. Draw or paint pictures of each part to create a class exhibition, so that visitors can walk their way through the story in the sequence of the artwork. You could even record sounds to be played with each picture.
- Role-play the hunter and his conversations with the blind man. Use a thought corridor (see p. 66) to explore what the hunter was thinking when he was tempted to swap his plain bird for the rainbow bird.
- Re-enact the story with sound effects. Perform your story to an audience and use the instruments you made in science lessons (see below) to provide the sound effects.

Explore the science that is directly linked to the stories:

- Hearing is one of our senses.
- We hear with our ears.
- Sounds get fainter as the distance from the sound source increases.

Explore the science that is indirectly linked to the stories:

- We have five senses: sight, hearing, touch, taste and smell.
- We can make sounds fainter by muffling them with certain materials.

Try these science activities:

SOUND AND HEARING

1.What can you hear?

Listen to the sounds around you in the classroom or outside. Which things make a sound? Are there sounds we can't hear? Use dog whistles etc.

2. Guess the sound

Listen to animal sounds (use internet clips) and try to guess which animal is making them. Hide objects in pots. Using only the sound made by the object when shaken, guess what is inside the pot. Talk about the sounds. Describe the sounds you can hear. Sort the pots into types of sound.

3. Make sounds

Make all kinds of instruments! You could make shakers and order them according to pitch or make elastic-band guitars – plastic boxes and baking tins work well. You could use bottles filled with water to tap or tubes of different lengths to whack. Keep talking about the sounds! You may even want to perform the story using the instruments for sound effects

4. Why is our ear that shape?

Look at human ears and animal ears. Roll up paper into cones. Use them as ear trumpets to hear better and to amplify voices. What does the cone shape do?

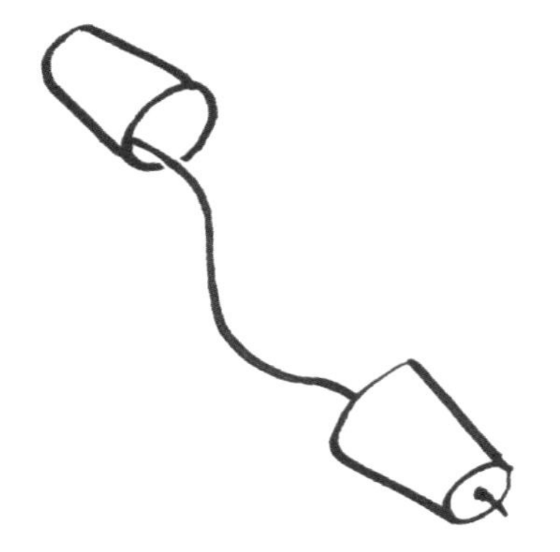

Observe how it amplifies the sound. Relate this to the shape of our ears and the way that animals will direct their ears to gather more sound. Make cup telephones and play with them. Demonstrate how your voice travels down a taught string but not a floppy one. Also show that rubbing the string makes very little noise until you add the cup on the end.

5. Make it quieter!

Read *Peace at Last* by Jill Murphy to set the scene for an investigation. Daddy Bear can't sleep because of the noise. Get a cooking timer/alarm clock that buzzes or beeps. Challenge the children to wrap up the timer or alarm clock with something that will muffle the sound most effectively to help Daddy Bear sleep. Give them all kinds of materials to choose from.

6. Sounds from far and near

Choose something that makes a small sound, such as a triangle. In a big quiet space, play the triangle behind the back of a child. When the child hears the triangle, they raise a hand. Then they take a step away from the triangle, and the triangle is played again. There will come a point where the child cannot hear the sound any longer. Measure the distance. This activity can be done as a demonstration first and then the children can try it for themselves in pairs, one playing a triangle for the other to hear. Investigate how far different sounds carry – they'll all have to be pretty quiet sounds, since a loud sound will carry across a school field.

Explore other curriculum areas that link to the story:

Reading

Read other sound stories. e.g. *Peace at Last* by Jill Murphy (Macmillan, London, 2013).

Fiction

Box up and write the story, or innovate using different animals caught in the traps or different dangers heard by the blind man.

Non-fiction

Explanation: Give a presentation to explain how the human ear works.

Explanation: Write an explanation about how the ears of animals are adapted to their habitat and position in the food chain (predators or prey).

Instructions: Write instructions on how to lay a trap.

Instructions: Write instructions on how to catch a bird. It could be a fantasy bird and, therefore, the way that it is caught could be fantastical too, e.g. 'Catching the Golden Trinket Bird' by leaving out a bowl of very heavy rings surrounded by low nets. The Golden Trinket Bird flies over the nets and picks up the rings from the bowl. As the bird tries to fly off over the nets the weight of the rings weighs the bird down, causing it to fly right into the nets.

Recount: Write about the hunting trip in role as the hunter or blind man.

Persuasion: Be the blind man and persuade the hunter to let you go hunting.

Information report: 'All about Ears' or 'All about Eyes'.

Music/Dance

Listen to music inspired by animals, e.g. *Peter and the Wolf* written by Sergei Prokofiev (1936) or *Carnival of the Animals'* composed by Camille Saint-Saëns (1886). Make up your own animal dance steps.

Design/Technology

Make animal traps! The simplest design is a pitfall trap – dig a hole in the earth and put a plastic cup into the hole so that the top of the cup is level with the ground. Insects and bugs will fall into the trap and won't be able to get out. Remember to check the trap daily, since the insects won't have food or water in there and will need to be released.

You could make more complex traps where a box or cage is propped up at one end by a stick. If the stick is pushed away the box falls on the animal. Test these with wind-up toy creatures rather than real ones to avoid injuring animals! You could even design and make fantasy traps for fantasy creatures.

Life Skills

What if you can't hear? How do deaf people speak to each other? Learn some sign language.

Some sounds warn us of danger. Find out about sounds that serve to warn us, like smoke alarms and sirens.

Topic 5.4 FORCES – PUSH, PULL AND TWIST

Story: The Wheel that Jack Built

Topic: Forces – push, pull and friction

Plot type: Quest

Genre: General fiction

This is a story about Clever Jack who finds a way to harness the force of the river by building a water wheel. The miller, the carpenter and the blacksmith all use the turning force of the water wheel to do their work for them until one day the wheel seizes up.

The science content in the story includes the forces of pushing and turning. It also introduces the idea of friction as a slowing force. It was inspired by a visit to Combe Mill in Oxfordshire, where a water wheel is used to power a whole barnful of machinery.

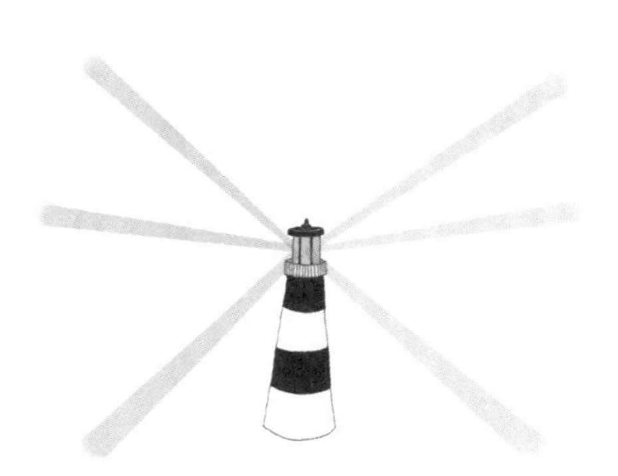

The Wheel that Jack Built

Once ... there were three friends: a miller, a carpenter and blacksmith. The miller lived by a fast-flowing river where he worked all day grinding wheat into flour by turning a millstone with two carthorses. The carpenter worked all day turning his lathe, by hand, to make chair legs and bedposts and spindles. The blacksmith worked pumping his bellows, by hand, to heat the forge to heat the iron.

One day, Clever Jack the farmer's son delivered some wheat to the miller. He watched as the miller ground it into wheat with the big millstone pulled around by the two horses.

'Hmm,' thought Clever Jack. 'I can think of a better way to do that!'

Jack fetched the carpenter and together they built a huge wooden wheel with a strong metal hub and lowered the bottom edge into the water. The wheel had paddles that hung into the river. The flowing water PUSHED the paddles and Jack's wheel turned. The wheel TURNED the central shaft and the shaft TURNED the heavy millstone. The miller poured in the wheat and the millstones ground it to flour.

'Not bad,' said Jack.

'Not bad at all,' said the miller. 'Much cheaper than horses. They eat all the wheat!'

The carpenter was thinking, 'If the force of the water can grind wheat, then why not use it to turn wood? It's much stronger than me.'

So they set to work again. When they had finished ... The water PUSHED the paddles and Jack's wheel turned. The wheel TURNED the central shaft and the shaft TURNED the lathe and the lathe started spinning. The carpenter started carving quicker than ever before on his new turning lathe. It was so much easier than working the lathe himself.

Jack was delighted. The miller was delighted. The carpenter was delighted.

Later that week, the blacksmith arrived with some hinges for the carpenter. He asked to see the water wheel that he'd heard so much about. When he saw the water PUSHING the wheel so powerfully, he started thinking, 'Hmmm, if the force of the river can turn a millstone and a lathe, then maybe it can help me. I could use that force to pump my bellows. It's much stronger than me!'

So they set to work again. When they had finished ... The water PUSHED the paddles and Jack's wheel turned. The wheel TURNED the central shaft and the shaft TURNED the crank. The crank PUMPED the bellows. The bellows blew air into the blacksmith's fire and made it really hot. The blacksmith heated his iron and hammered it until he had made a beautiful new handle for the barn door.

The blacksmith was delighted. The Miller was delighted. Jack was delighted.

After a few months, the miller noticed that the millstone wasn't turning so fast. The carpenter's lathe had also slowed down and the blacksmith's bellows weren't pumping fast enough to keep the fire hot. They all went to look at the water wheel. It had slowed right down and was making a grinding noise. It was also looking rather hot; a thin wisp of smoke spiralled upwards from the centre of the wheel.

'Oh dear,' said the miller.

'Oh dear,' said the carpenter.

'I'll fetch Jack,' said the blacksmith. 'He'll know what to do.'

'Don't worry!' said Jack. 'It's just rust on the wheel hub. It's making the surfaces rough so the wheel can't turn.'

He poured some oil on the centre of the water wheel where the wheel hub turned on its fixings. Slowly but surely, surely but slowly, the water PUSHED the paddles and the water wheel turned, faster and faster until flour poured from the millstones, the lathe turned like a spinning top and the blacksmith's fire was hotter than hot.

As for Clever Jack, he started a company called Clever Jack's Water Wheels, made a good living and moved to a fine house on the hill.

And all from the wheel that Jack built.

Top tips for telling

This story is all about imagining the way that the water wheel powers the grinding, the turning and the pumping machines. To do this, the class have to be able to imagine what is going on, so they need to know about millstones, lathes and forges. You may like to show pictures of millstones, a lathe and bellows as you tell the story, and then explain afterwards how the machines work. Or you may want to explain how millstones, lathes and bellows work before you begin the story. The water wheel can be explained as part of the story, since it is easier to understand.

When you describe what the water wheel does, emphasise the doing words (PUSHED, TURNED, PUMPED) and use actions to emphasise the movements. These sections can be told with a rhythm and the children may like to join in. The final moment of drama is when the miller's wheel starts to slow down and heat up. Draw out the uncertainty of this moment before Clever Jack solves the problem with his can of oil!

Ways to work with the story

Teach the story using HMSS

Deepen the story with activities such as these:

- Re-enact the story with children playing the part of the machines and showing the forces at work.
- Role-play Clever Jack explaining to the miller and carpenter how the water wheel can be used to work the machines.
- Role-play Jack explaining to the others how friction has slowed the wheel down and how the oil will reduce the friction.
- Explore the movements of the river and the machines in mime, dance or art.
- Play with toys with visible cogs. Watch how one cog turns another.

Explore the science that is directly linked to the story:

- Push, pull and twist are all forces.
- Forces can make things begin to move (if they are not already moving), make them stop and change their speed or shape.
- Friction affects the way two surfaces move over one another.

Explore the science that is indirectly linked to the story:

- two forces can act in opposition to one another;
- ways to reduce slowing forces, e.g. streamlining;
- gears and levers allow a smaller force to have a greater effect.

Try these science activities:

FORCES – PUSH, PULL AND TWIST

1. Finding forces in the classroom

Investigate toys that use springs, clockwork, or magnets, push-along toys, pop-ups and cars on ramps. Note that forces speed things up and slow things down.

2. Measure pushes and pulls

Use bathroom scales and newton meters to compare pushes and pulls. Find out that forces are measured in newtons. Make your own force measurer by hanging 100 g masses on an elastic band and calibrating it by noting how far it stretches with each additional 100 g mass. (Gravity pulls down on a 100 g mass with 1 newton of force.)

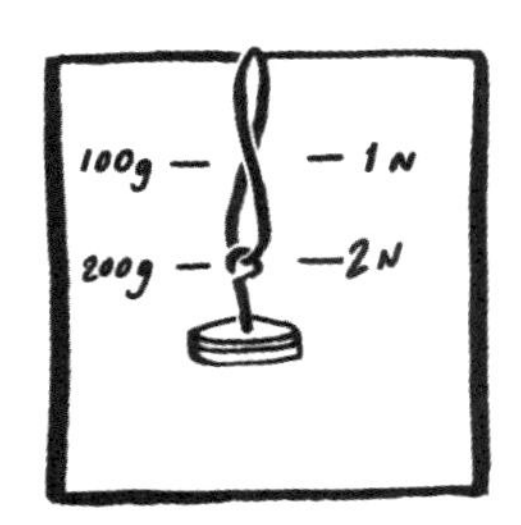

3. Measure friction

Use newton meters to find out which shoe has the best grip. Investigate which shoe is hardest to move across a table top. Is it the tread, the size or the material that makes it more grippy? Measure the friction by pulling the shoe horizontally across the desk using the newton meter.

4. Investigating forces that act in different directions

Investigate cars rolling down slopes onto different surfaces – gravity acting in opposition to friction. Find out which slopes have the greatest friction by comparing how far or how fast the cars travel on the surface at the base of the slope. You could compare carpet and wood, since these produce different amounts of friction. Draw diagrams of the forces to show which way they are acting.

5. Marble run

Make a marble run on a piece of board. Support one end of the board to make a ramp and design a course with a limited number of lolly sticks and Lego bricks for the marble to travel down. You could challenge the children to use their knowledge of friction to make the slowest marble run possible with the equipment you provide. Give them just enough adhesive putty to stick the lolly sticks to the board. NB: The marble must move – if it stops then the marble run does not meet the challenge and must be adjusted to let the marble move.

6. Streamlining

Investigate the slowing forces in water – water resistance. Look at fish, boats and penguins to see how they are streamlined to reduce resistance. You could even make boats with hulls of different shapes and race them to see the different resistance.

Explore other curriculum areas that link to the story:

Reading

Read other stories about machines e.g. *Clockwork* by Philip Pullman (Yearling Books, New York, 2004) or the imaginary machines in the stories about Mrs Armitage by Quentin Blake (Red Fox, London).

Fiction

Box up and write the story, or innovate using new machines to link to the water wheel. These could be realistic or fictional machines.

Poetry

Perform poems with a similar style to the story, such as *The King's Breakfast* by A.A. Milne (from *The Complete Poems of Winnie-the-Pooh* [Dutton, New York, 1998]).

Non-fiction

Persuasion: Prepare a sales pitch or advert for a water wheel.

Explanation: Write a page from a manual explaining why the wheel must be oiled regularly.

Instructions: Write instructions on how to use and take care of the water wheel. This could be an owner's manual or video.

Discussion: Discuss the pros and cons of replacing cattle power with water-wheel power.

Information report: Write a report about water wheels or a local mill.

Information report: Write a report entitled 'All about Blacksmiths'.

Design/Technology

Make a moving model. This could be a balloon car in which an inflated balloon pushes a model car along. You can make one by taping two drinking straws to the base of a box and putting thin dowel axles through the straws.

Fix wheels to the dowel and make sure they turn freely when the dowel spins inside the straw. Then fix a balloon to the car and push a short length of straw into the end of the balloon so that the straw points to the rear of the car. When the inflated balloon is released the jet of air should push the car along. It works best on a very smooth floor with a car made from very lightweight materials.

History

Find out about Newton and how he came up with the idea of gravity after watching apples falling from the trees.

Find out about other simple machines such as the Archimedes screw or the shaduf in Ancient Egypt.

Topic 5.5 FORCES – MAGNETISM

Story: The Magic Stone

Topic: Magnetism
Plot type: Quest
Genre: General fiction

This is a story about a lodestone – a naturally occurring magnet. The lodestone is used by a boy to find his way home after playing in the forest. He pretends he is using magic to find his way but his sister becomes suspicious and eventually works out how to use the lodestone for herself.

The science content of the story includes the attraction of iron to a magnet and the use of magnets to make compasses.

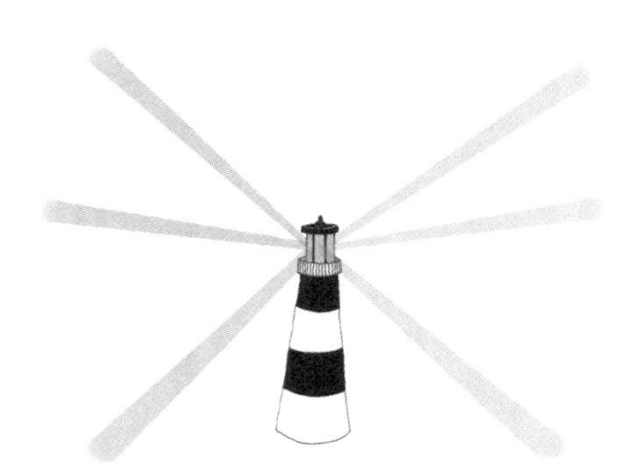

The Magic Stone

Once ... long ago, there lived a girl named Erika. She lived on the edge of a huge forest. Every day, her brother went off to school. Erika stayed at home with her mother, sewing and cooking. But she hated sewing and was terrible at cooking.

'O Erika, not another knot ... O Erika, please don't add any more salt ... O Erika, watch that pot –it's burning!' her mother would cry.

And Erika would slam it all down and storm out of the house.

In the evenings, she loved to go out walking in the forest to the south of the village with her brother. They would wander here and there until Erika had no idea where she was or which way was home. She loved the shade of the forest and the feel of the dust between her toes. She loved climbing trees and watching the animals. Sometimes at dusk she heard wolves howling and she was glad that her brother was there to find the way home. At night the forest was dangerous!

Luckily, whenever it was time to go home, her brother always seemed to know the way. He would just walk in a straight line through the dense trees, always in the right direction.

One day she asked him, 'Leo, how are you so good at finding your way?'

'I'm not really your brother,' he said, ruffling her hair. 'I'm a bird in the body of a boy and you know that birds can always find their way home.'

Erika shook her head. 'Leo, you can't be a bird – you are terrible at singing.'

'Ha,' he said.' 'Alright, I have an amazing sense of smell and I can smell our mother's cooking from miles away.'

Erika tried sniffing the air. Leo laughed. Erika blushed and felt silly.

'Alright,' he said, 'I'll tell you the truth. I have a magic stone – here around my neck.' He waved around the ugly lump of rock he always wore on a string.

'Shut up!' she snapped. 'Don't make fun of me.' Erika stamped her feet and decided that she would learn how he did it. She would watch him carefully and find out his secret.

The next day, as evening fell in the forest, she watched her brother carefully. He stood with his back to her, clearly up to something. Maybe he can smell mother's cooking, she thought. Erika crept up behind Leo – and saw him take off the stone necklace and let it hang by the string so that the stone swung this way and that.

'Leo ... is that a magic stone? Are you a magician?' she asked, her eyes wide.

'Yes it's magic! It can even pull the nails right out of your sandals!'

Erika glared at him, pink in the face and wild about the eyes. 'Stop making fun of me!' she said.

Leo laughed. 'You wouldn't understand. You're just a girl.'

Well, that made Erika mad. Mad and determined. It had something to do with the stone and if Leo wouldn't tell her then she would find out for herself.

The next day, while the boys were at school, Erika slipped into Leo's room and took the stone necklace from the shelf and put it over her head. She went to the forest to the south of the village and wandered through the trees until she could no longer see her house. Then she took off the stone necklace and looked at it. She held it tight and she wished, 'Show me the way. Show me the way.' But it gave her no sign.

She rubbed it and it did nothing. She held it to her ear but she could hear no sound. Time passed. She couldn't get it to do anything. She threw it to the floor and sat down. She tapped it this way and that with her foot. And then something strange happened. It began to follow her sandal. If she moved her sandal slowly away, the stone came too, sticking to the tacks in the sole. A bit like what Leo had said ...

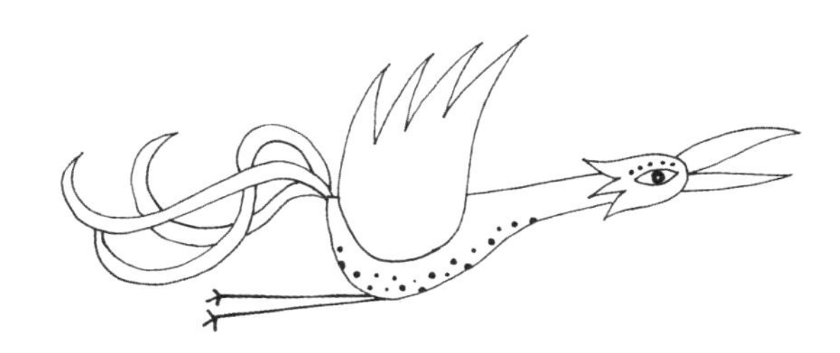

She jumped up. It was magic! She was afraid to touch it again. All of a sudden she felt terribly afraid and terribly alone. She knew she had to go back. Holding it as far away from herself as she could in case it did any other strange tricks, she picked up the stone by the string. And as she watched, the stone spun to the left and to the right ... and to the left and to the right ... and then it settled and stopped moving. Erika was curious and gave it a tap. She waited while it spun left then right and stopped again in exactly the same place. There was a mark on one end of the stone.

'That's how he did it!' she thought. And she fixed her eyes in the direction of the mark and followed it. It led her straight home.

The next day, when she was out with her brother she said, 'I will lead us home today.'

Leo laughed and swung down from the trees to watch. She took the stone necklace from Leo's neck and held it out. She waited for the stone to be still and she pointed in the direction of the mark.

'I know the magic!' she said.

Leo smiled and said, 'What a clever sister I have. This is a lodestone. It will always point north. It's not magic. It's just a special stone.' And he took the stone and broke it in half.

'Oh Leo, you've broken it!' Erika cried.

'No,' he said, ruffling her hair, 'it will still work; I've just made it into two!'

He gave her the half with the mark. She put the stone on a string and wore it around her neck. She didn't think it was ugly any more. From that day on, she could always find her way home from the forest. As she grew older, she would go deeper and deeper into the forest to the south of the village, finding pools and caves and waterfalls. Wherever she went, her lodestone helped her find the way home before nightfall.

How do you think the stone worked?

Top tips for telling

In this story, Erika learns the science behind the 'magic'. Evoke that possibility in the telling to build up the suspense and then her amazement at discovering the powers of the stone. Also, make sure you note the possible dangers of the forest, emphasising the importance of getting home safely. The rest is about the two characters: Erika is curious and determined, especially when teased by her brother. Leo is playful and loving. His gift at the end embodies this loving quality.

Ways to work with the story

Teach the story using HMSS

Deepen the story with activities such as these:

- Explore the idea of forces by acting out pushing and pulling, attracting and repelling.
- Re-enact the part of the story where Erika is trying to work out how her brother knows the way home.
- Role-play the scene when Erika finds that the lodestone is attracted to the nails in her sandals. Explore how her initial fright becomes curiosity and then discovery.
- Role-play Erika telling her mother about how she worked out the way to get home.
- Explore Erika's dilemma – should she pick up the lodestone now she thinks that it is magic? This could be done using a thought corridor (see p. 66).

Explore the science that is directly linked to the story:

- Magnets attract/repel each other depending on which poles are facing and this is an example of a force.
- Magnets will attract some materials and not others.
- Magnets will align themselves with the poles of the Earth if allowed to move freely.

Try these science activities:

FORCES – MAGNETISM

1. The discovery lesson

Allow the children to play with magnets. Include a variety of shapes such as ring magnets. Let them wonder at one magnet pulling or turning another and the feel of two magnets repelling. Ask questions: Will it stick to any material? Will it work through the table? Does it attract at both ends? Some magnets are colour coded, not labelled – which colour is north? Draw diagrams of what happened. Label the north and south poles.

2. Sorting materials

Investigate which materials will stick to a magnet. Begin with a mixture of materials and then move on to testing only metals, noting that there are very few magnetic metals. Record this in a sorting diagram. Sort fizzy drink cans using a magnet – can you find the aluminium ones? This is how steel cans are separated in the recycling industry.

3. Magnets in real life

Look for magnets in home or school – this could be done as homework. Find magnets in cupboard door holders, fridge magnets, bag clasps, etc. Record them as pictures or photos. Make a magnet by stroking a strong magnet along a paper clip in one direction for a few minutes. Watch it attract other paperclips.

4. Measuring the strength of a magnet

Begin by giving the children a selection of magnets of different strengths and ask the children to prove which one is the strongest by testing them in some way. Share all their methods. If they are fair, allow the children to gather data that way. If they don't come up with a fair test, use this one: lay a paper clip alongside the zero mark on a ruler, lay the magnet alongside the ruler and move it slowly towards the magnet, noting the distance when the clip jumps.

Draw bar charts and explain which magnet is the strongest and the evidence for this.

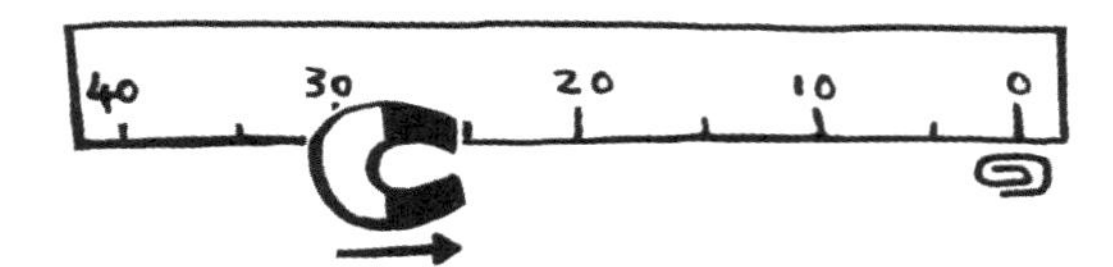

5. Using magnets as compasses

Research how magnets are used in compasses and how the Earth has a magnetic field along which the north and south poles of a magnet will align if allowed to move freely. Magnetise some needles (by stroking them as before) and float them on tiny cork rafts to make compasses. Observe how many times out of ten the needle will float into a roughly north–south alignment. Relate all this to the lodestone in the story.

6. Discover how magnetic toys work

Play with fishing games, levitating toys (containing ring magnets) and 'magic' games where a magnet is used to 'magically' move an object. Decide how you might use a magnet in your own toy. Design your own toy to build in Design/Technology.

Explore other curriculum areas that link to the story:

Fiction

Box up and write the story, or first innovate by setting the story in a new landscape (ocean, lake, caves, north pole). The children may think of other objects in that setting which might be made of iron and attracted to the lodestone.

Non-fiction

Instructions: Write instructions on how to build a magnetic toy.

Instructions: Write instructions on how to use a compass and map to navigate.

Information report: Give a presentation on the different kinds of magnets.

Persuasion: Create an advert for a lodestone or compass.

Explanation: Write an explanation of how a compass works.

Explanation: Be Erika and explain how you found your way home.

Recount: Be Erika and tell a friend your story.

History

Research the invention of the compass and how this enabled ancient peoples to explore the world. You could also investigate other means of navigation, like the position of the stars or the angle of the sun.

Maths

Learn to draw sorting diagrams and use them to sort objects according to their properties. Draw a Venn diagram of objects attracted to a magnet and those which are not. Some objects are made from more than one material (e.g. plastic-handled steel scissors) and could be placed in the overlapping section of two circles in the Venn diagram.

Use the data collected in science activity 4 (below) to draw bar charts comparing the strength of different magnets.

Geography

Learn to use compass directions in relation to maps and given directions.

Design/Technology

Make magnetic toys of the children's own design based on toys seen in science lessons.

Topic 5.6 LIGHT

Story: The Torch

Topic: Light travelling in straight lines, light sources and reflective materials
Plot type: Overcoming the monster
Genre: General fiction

This is a story about a young man who is sent away to war. He has taken a torch with him. When he is captured by the enemy, he uses the torch to help dig a straight tunnel and escape to a rendezvous point where he must signal to the rescue plane. When his torch runs out at this crucial moment, and in the absence of any other light source, he uses the reflector inside the torch to signal to the plane by reflecting the moonlight.

The first scientific concept covered in the story is the idea that light travels in straight lines, so Tom can only see down the tunnel if it is straight. The story also involves reflection of light from shiny surfaces and mentions various other light sources.

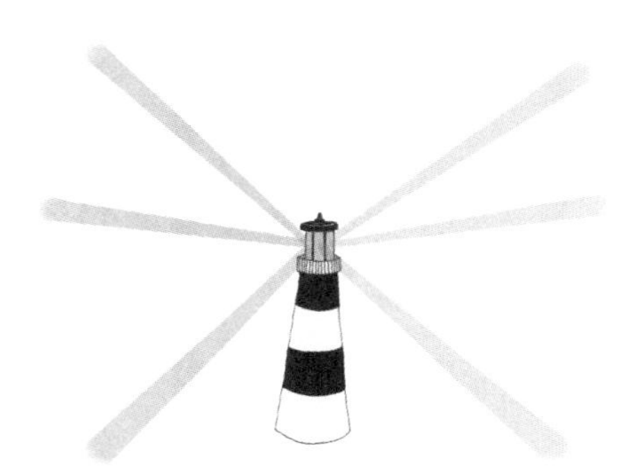

The Torch

Once ... there was a boy called Tom who lived in house in a little village with his mum and dad.

When Tom was eleven, his mum and dad gave him a torch for Christmas. Tom was so pleased with it. At night, he'd go walking in the woods with his torch, shining it to find his way. When he put a tent in the garden, he stayed up all night, reading with his torch. One day, in the woods, he found a little cave and spent all day there looking in all the nooks and crannies. He felt like an explorer! He loved that torch! It was always there, tucked away in his pocket, just in case!

When Tom was sixteen, the Second World War broke out and Tom joined the army to fight for his country. Before long, they sent him to fight on the front line and, as always, he took his trusty torch with him. When he was crouching next to the guns, he always had it with him. Just the feel of it at his side made him feel closer to home.

One dreadful day, Tom's group was surrounded by German soldiers.

'Surrender!' they shouted.

Tom had to lay down his weapon and he was taken captive. He was a prisoner of war. As soon as he arrived at the camp, the guards stripped him of all his possessions. They took everything away, including his torch! At night, when Tom was trying to sleep, he wished he had it back, as it was like a little piece of home.

One morning, at roll-call, Tom saw his torch sticking out of the pocket of the Camp Captain's coat. He couldn't take his eyes off it. He was so pleased to see it and furious that it had been taken from him. Every time the Camp Captain was in sight Tom watched him, looking for the torch. It was often in sight. Then one warm sunny morning, Tom saw that the Camp Captain had left his

P
PHOTOCOPIABLE FOR USE IN SCHOOL

coat unattended on the seat of his jeep. Tom didn't waste a second. Checking for guards, he walked up to the jeep. He only had a moment before the Camp Captain came back. His heart was beating like a drum as he slipped his hand into the pocket of the coat. It was there! He felt such excitement at the familiar shape of the torch in his hand. Quickly shoving it into his pocket, he moved away from the jeep before he was spotted. He looked all around. No one had noticed. He let out a silent sigh and went back to his bunk.

Then, one day, a new group of prisoners arrived at the camp. Their captain was young and full of energy and he talked endlessly about escape. 'It's our duty to escape!' he said. 'We must dig a secret tunnel under the fence and out into the forest.

' Tom was delighted to have something to do at last. He was the smallest, so they asked him to help dig the tunnels. They had to dig in a straight line from the hut, under the wall and out into a clump of trees where they would be hidden from the guards. The tunnel had to be straight so they would come out in the right place.

'How will we know if the tunnel is going the right way?' asked Tom.

'If we only had a torch,' sighed the Captain. 'We could shine it down the tunnel. Light goes in a straight line. You'd know the tunnel was straight if you could see the torch from the other end.'

Tom's hand closed around the one thing he had that reminded him of home. The one thing he had in the whole of the war that was actually his. He didn't want to give it up. But he did want to escape and his torch might be the answer. Reluctantly, he held it out for the Captain.

They couldn't use the torch very often. They had to save the batteries. But each evening Tom would check that his light beam travelled straight down the centre of the tunnel. It took several months to make the tunnel long enough to take them beyond the fence and out to the clump of trees.

The Captain used the time to make plans. He knew a place where a rescue plane would pass, just before dawn, on the first day of every month. If they could reach the spot and use the torch to signal to the plane, then it would land and take them home.

Time passed slowly. At last, it was the last day of the month. They had to leave that night. Everyone assembled in the hut and the Captain sent Tom ahead to dig the last part of the tunnel, up to the surface. Tom dug his way up and out, right in the centre of the clump of trees and out of sight of the guards. 'Thank goodness for my torch,' he said under his breath!

The Captain came after him, then one by one, the rest of the men followed. It was a long hike to the rendezvous and they had to wait in dark, shady places until a cloud covered the bright, full moon and provided them with a shadow; they couldn't just walk across the open spaces in the moonlight. But, at last, they reached the spot. The Captain called for Tom to come with his torch. The Captain flicked it on ready to flash at the plane. And ... nothing ... The batteries had run out. They shook it and tapped it. It gave a final yellow flicker and went out again.

Tom and the Captain looked at each other. They could hear engines in the distance. They had come so far and the plane was coming – all they needed was a light to signal with. The Captain, usually so full of ideas, looked lost. 'We need a light!' he whispered. 'How can we find a light?' Tom tried to think. Nobody had batteries or matches. What could they do?

Then, he had an idea. He unscrewed the torch and took out the metal reflector from behind the bulb and polished it on his shirt until it was gleaming. Then he stepped out into the full glare of the moonlight, held up the reflector to the moon and moved it until it caught the light. He turned the reflector this way and that, doing his best to flash the signal as the plane drew closer. He directed the reflected light right at the cockpit of the plane. The men watched and held their breath. And with a dip of the wing tips, the plane flew over ... turned ... and came back ... descending and ready to land. The men grinned with gladness.

Back home, Tom went to see his mum and dad and told them the story. 'If it wasn't for this torch,' he said, 'I might have been caught and maybe shot for escaping. This torch is the best present I have ever had. It saved my life! Thank you!'

Top tips for telling

First evoke all the enthusiasm of a boy for his treasured torch. Then emphasise the tension of the three dramas: getting the torch back, tunnelling, and reflecting the moonlight.

You need to evoke the fears and uncertainties of each situation then the relief of resolution. Linger a little on the part about how the torch helps Tom to dig the tunnel straight, so it is clear.

Ways to work with the story

Teach the story using HMSS

Deepen the story with activities such as these:

- Explore the dilemma faced by Tom when he is asked to lend his torch to the Captain. Use a thought corridor (see p. 66) to suggest what he is thinking at that moment.
- Re-enact the moment when Tom uses the reflector to signal to the plane. Tom could explain to the Captain what he is doing.
- Role-play the Captain explaining how to dig a tunnel and use the torch to check that it is straight.
- Draw pictures or diagrams of the torch and tunnel or the moon, mirror and plane, showing the path of the light as it reflects off the moon onto the reflector and then into the eyes of the pilot.

Explore the science that is directly linked to the story:

- Light comes from a light source.
- Light can be reflected from some surfaces.
- Light appears to travel in straight lines.

Explore the science that is indirectly linked to the story:

- We need light in order to see things, and dark is the absence of light.
- We see things because light travels into our eyes.
- Light from the sun can be dangerous and there are ways to protect our eyes.
- Shadows are formed when the light from a light source is blocked.
- The moon is not a light source – it reflects the light of the sun.

Try these science activities:

LIGHT

1. Light source or reflector?

Sort light sources from things that merely reflect light. One way is to make a dark box (a completely dark shoebox with a peephole) and put items into it. Shiny things are not visible in the dark but light sources are. Make sure that the children understand that the moon is not a source of light but reflects the light from the sun.

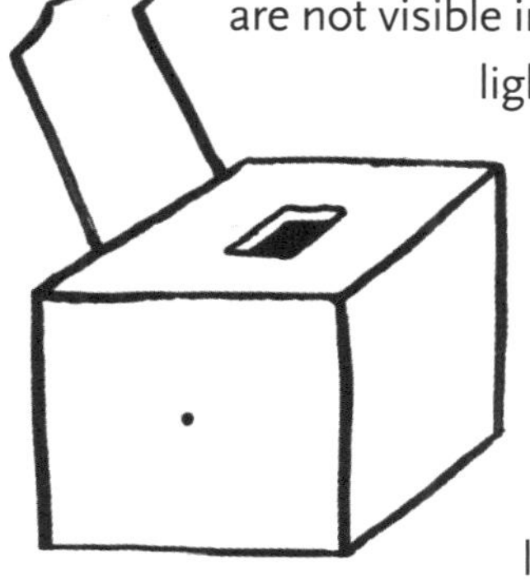

2. Finding the best reflector

Compare materials to see which reflect light. This is the moment to let the children use their watches and rulers to make patches of reflected sunlight on the ceiling. If it's overcast and you can dim the room, you can use torch beams to reflect patches of light onto the walls. You'll need to hold the torch close to the ruler and reflect the light onto a wall less than a metre away. Relate this back to the story – what else could Tom have used to reflect the light? You can also use a narrow beam of light from a light box and mirrors in a dark space (under a table covered with blackout material) to show

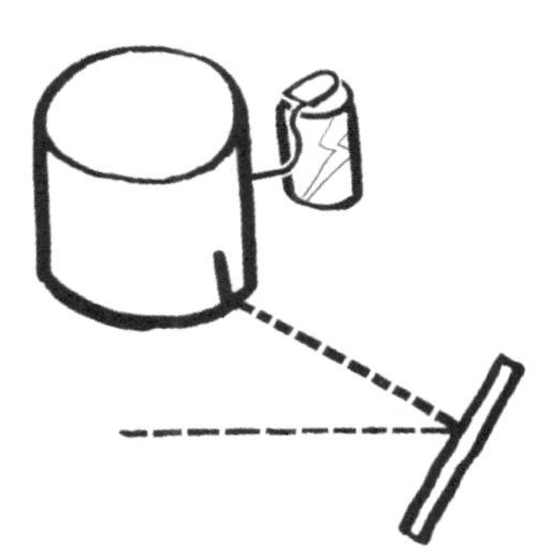

the light bouncing off the mirrors at predictable angles. Get the beam of light travelling along the floor and then hold the mirrors in the path of the light, perpendicular to the floor, to reflect the light beam at an angle.

Use this knowledge to make a simple model periscope. You can find a simple template on the internet.

3. Shadows

Investigate the effect on shadows when you move the object or the light source and the effect of changing the transparency of the object making the shadow. This can all be done by making shadow puppets from a variety of materials and creating a show in front of a projector on a blank screen. You could make shadow puppets of the characters in 'The Torch' and perform the story.

4. Modelling and drawing light

Draw diagrams to show how shadows are made, demonstrating that light travels in straight lines. Use computer-modelling programs to play with lights and mirrors. It can be hard to find an area that can be blacked out sufficiently to see rays of light bouncing off mirrors, so computer modelling offers a whole-class option for investigating the reflection of light beams. There are some great resources on the BBC Bitesize website (www.bbc.co.uk/education).

Continued overleaf

LIGHT *continued*

5. The eye

Learn about the structure of the eye and how light enters the eye through the pupil so that we can see. Learn about protecting our eyes from the damaging effects of the sun and other hazards such as laser pens. You could even make a simple pinhole camera from a shoebox to show how the image forms on the back of the eyeball upside down. You can find plenty of detailed instructions on how to make pinhole cameras on the internet.

6. Research the many uses of light

Find out about the use of light in modern technology, such as lasers in CD/DVD players, optic fibres for internet and telephone signals, etc. With older children you may want to introduce the idea that light is reflected from a mirror at the same angle at which it hits the mirror, i.e. the angle of incidence is equal to the angle of reflection. This is an extension of activity 2 above. Measure the angles made by the beam of light entering and reflecting off the mirror. If you lay paper on the floor and make the beams of light travel over it, you can trace the paths of the light with a pencil and then use a protractor to measure the angles.

7. Properties of materials

You may choose to include a section of the materials topic and test materials for transparency.

Explore other curriculum areas that link to the story:

Reading

Read other wartime stories, e.g. *When Hitler Stole Pink Rabbit* by Judith Kerr (HarperCollins, London, 2008).

Fiction

Box up and write the story, or innovate by resetting the story in another place where you could be trapped, e.g. a boarding school, a kidnapper's lair or a dragon's cave, using the same escape method. Alternatively, innovate by thinking of other ways of using light to escape, e.g. Morse code, finding the way at night, laser signals or even blinding the enemy with bright light.

Non-fiction

Instructions: Write instructions with diagrams on how to dig an escape tunnel.

Instructions: Write instructions on how to signal using a mirror and reflected light.

Information report: Write a report entitled 'All about Electric Lights – from Lighthouses to LEDs'.

Information report: Prepare a presentation on ways to send messages – from Morse code and semaphore to the use of cyphers and invisible ink.

Recount: Write an interview with Tom for a newspaper or TV report.

Recount: Write Tom's diary entry for the day of the escape.

Persuasion: Be the Captain and persuade a reluctant Tom to help.

Discussion: Discuss whether Tom should try to escape.

History

Study World War II, including escape stories.

Find out about the use of periscopes by submarines.

Find out about the persecution of the Jews during the Holocaust through the story of Anne Frank or by reading *Rose Blanche* by Ian McEwan (Red Fox, London, 2004).

Design/Technology

Look at torch designs and draw diagrams. Practise building the circuits required to make a bulb go on and off with a switch. Use electrical components to design and build a torch with a switch.

Religious Education

Find out how the festivals of light such as Divali, Christmas and Hanukkah are celebrated. Hold your own celebration so the children can experience it for themselves.

Topic 5.7 ELECTRICITY – SERIES CIRCUITS

Story: The Lighthouse Keeper's Son

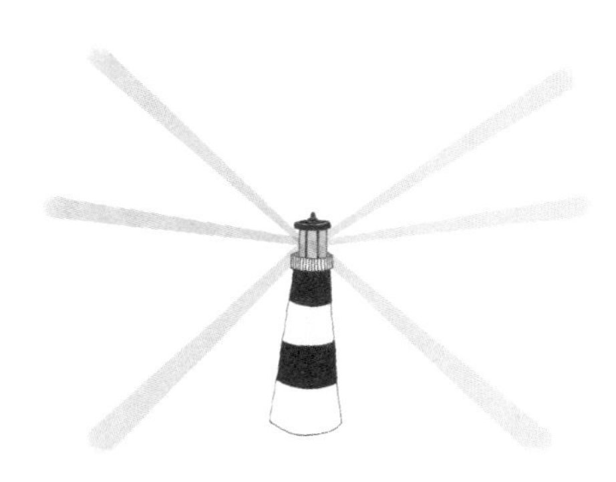

Topic: Simple circuits, switches and conductors
(Could be extended to include parallel circuits)
Plot type: Quest
Genre: General fiction

This is a story about a lighthouse keeper's book-loving son who decides to link all the lighthouses on an island into one electrical circuit so they can all be operated by one switch. Now he can run all three lighthouses on his own. When a big storm pulls a wire loose, all the lights go off and he has to solve the problem by himself.

The science in the story includes the fact that all the components in a circuit go off when a circuit is broken by the switch or a broken wire, showing that circuits must be complete for electricity to flow. There is an opportunity to test for electrical conductance when you explore the part about trying to fix the broken wire in the storm. The story can be resolved (using the first ending) with three separate series circuits and three generators, or (using the alternative ending) with three lamps in parallel and one generator, should you want to take the science that far.

N.B. There are, of course, other ways to wire the circuit but they are not included here. The story aims first to introduce the simplest circuit, like the one the children will make in the classroom. Once they have learned that a circuit must be complete for electricity to flow, you may want to address other ways to wire the circuit, such as the one included in the alternative ending to the story.

The Lighthouse Keeper's Son

Once ... there was an island far out to sea where the wind blows hard and the wind blows fierce and the ships get blown right up onto the rocky shore. On this island lived a lighthouse keeper who had three sons. Every night, two of his sons would follow him up the stairs. 'Let me light the lamp, Father!' they would cry... but not Hamish, who had his head in a book.

The father showed his sons how to polish the lens and light the wick to burn the oil and make the beacon shine out for the ships at sea. Hamish wasn't interested. Hamish just wanted to read. 'What use are books out here lad?' his father would say.

Time passed and the sons grew up. The two older boys wanted to be lighthouse keepers like their father. There were two other lighthouses on the island and one by one the sons moved out and took over each of them, leaving only Hamish at home.

'Time to light the lamp, son!' the father would say as he disappeared up the stairs. He always hoped that Hamish would follow. But Hamish never did. 'Why won't he take an interest?' the lighthouse keeper thought. 'Doesn't he care?'

It made the lighthouse keeper angrier and angrier until one day he lost his temper. 'Why can't you be more like your brothers? When are you going to get your head out of those books and get a proper job? When are you going to DO something with your life?' He grabbed the book out of Hamish's hand and threw it into the fire.

'NOOOO!' shouted Hamish as he jumped out of his comfy chair to fish the book out of the flames. But the book was ruined. From that day on they barely said a word to each other.

Time passed. The lighthouse keeper got older and wearier. Then one day the two older brothers were called up to join the army and fight in the war. The lighthouse keeper was distraught. He couldn't run three lighthouses on his own! How could he turn all of them on at lighting-up time?

Hamish said, 'Dad ... I've been thinking about the lighthouse. Can I show you something in my new book?'

'Lad, books won't help light the lamps,' the lighthouse keeper said and he turned to go.

But Hamish carried on. 'Dad ... there's something new on the mainland. It's called electricity. I think we could use it to light the lighthouses. Not just one of them ... all of them.'

The lighthouse keeper stopped.

'Electricity can light up a huge bulb. It flows from a generator, down metal wires, and when it flows through the light bulb it makes a tiny wire inside get so hot that it glows. It gets hotter and hotter and makes a really bright light.'

The lighthouse keeper looked interested.

Hamish carried on: 'It burns clean – it won't get sooty.'

'Clean?' said the lighthouse keeper. 'I won't have to clean the lens?'

'No, Dad, you won't have to clean it,' said Hamish, 'and if we wire all the lighthouses together in one loop we can turn them all on with one switch. Imagine that, Dad. No more climbing the stairs. No more cleaning the lens. Just a flick of a switch.'

'One switch? Hmmm,' said the lighthouse keeper. 'Maybe your books are useful after all.' He asked a lot of questions and Hamish bought more books to find the answers. But eventually his father agreed. 'Son, I think we should give it a go.'

The workmen came over from the mainland with three electric lamps. They put huge cables in a big circle, supported by sturdy pylons, connecting up all three lighthouses on the island to a brand new generator. At the top of each lighthouse they put a fine new electric lamp. And in the lighthouse keeper's kitchen they placed a huge electric switch. When everything was connected into one huge loop, the lighthouse keeper threw the switch. Everyone held their breath. And then, as one, the three lamps flickered on.

For the first time in his life, the lighthouse keeper looked at his son with pride.

'Well done, son!' he said.

Time passed. Hamish took over the three lighthouses and his father retired. As he left, he took the cellar key from his pocket. 'Just in case!' he said and he hung the key by the door.

Hamish was quite happy. He read all day in the peace and quiet and he flicked the switch each day at dusk and dawn. And the ships saw the lights and stayed safely out of the way of the rocks.

Then came a day where the wind blew hard and the wind blew fierce and Hamish could see there was quite a storm brewing out at sea. The wind made all kinds of noises, whistling and growling, creaking and groaning. And then it began to make a new noise – a grinding noise. It didn't sound right. Hamish put on his coat and went out into the storm.

PHOTOCOPIABLE FOR USE IN SCHOOL
P

The wind whipped his coat about and the rain lashed his face. The cables were twisting in the wind, swinging one way then the other. Then, all of a sudden, the lights went out. All three of them. A cable had come right off its fixings and was swishing about like a snake.

All the ships at sea began to wail. They honked their horns to tell Hamish the lights had gone out. They honked their horns to tell the other ships to keep out of the way and they honked their horns in terror of being washed up on the rocky island shore.

'NOOOO!' screamed Hamish. And still the ships wailed in the dark. Hamish didn't know what to do.

As he stepped inside, he saw the cellar key hanging on the hook. 'Just in case,' his father had said. He took the key and went down to the cellar. There, at the back, in a neat line were three clean oil lamps.

'O Dad! Thank you,' he breathed as he heaved the first lamp up the stairs.

He lit the lamp as fast as he was able. And the ships sounded their horns in thanks. He loaded the other lamps into his van and went to each lighthouse until all the lamps were lit and all the ships were safe.

When the workmen came from the mainland, the following week, to fix the cable, Hamish had some new ideas. It would be more work but he knew it was the right thing to do. He had each lighthouse linked up to its own generator on its own loop of wire. Now there were three switches to turn off at dawn and on again at dusk, but Hamish didn't mind. At least he didn't have to climb the stairs and polish the lenses! The ships were safe and the job was done properly. And the three oil lamps stayed, polished and clean, in the cellar. And the key hung by the door, always.

Just in case.

If you want to teach about parallel circuits you might use this ending instead, starting at 

When the workmen came from the mainland, the following week to fix the cable, Hamish had some new ideas. It would be more work but he knew it was the right thing to do. He moved the generator to the centre of the island and he had each lighthouse linked up to the generator on its own loop of wire. Now there were three different switches to turn off at dawn and on again at dusk, but if one of the wires broke, only one of the lighthouses would go off. The other two would be fine, keeping the ships safe. The three oil lamps stayed, polished and clean, in the cellar. And the key hung by the door, always.

Just in case.

Top tips for telling

The heart of this story is the relationship between Hamish and his dad. Make sure you convey the different frustrations of father and son and then the surprise and pride when the electric idea works.

It is also important for the children to understand that lighthouses are a matter of life and death. Make sure you evoke that when setting the scene, so that they realise why the lamps matter so much.

Description is important in this story. Your listeners need to picture the lighthouses with the old oil lamps, and then the new electric lamps linked in a circle by wires moving in the wind. Make that clear, especially the circuit, so your audience fully imagines the lights, the wires, the generator and the switch.

Ways to work with the story

Teach the story using HMSS

Deepen the story with activities such as these:

- Explore the movement of electricity flowing around a circuit. You can use a loop of rope and allow the children to hold it as they stand in a circle. They all have to pass together to move the rope around the circle just as electricity is flowing at all points in the circuit at the same time.
- Re-enact the workmen building the circuits around the island, or the wires being blown off their fixings in the storm. You could use real wires, batteries and bulbs on a model island.
- Re-enact in drama, mime or dance the workmen building the circuits and getting the electricity flowing.
- Role-play Hamish explaining to his father how to wire up the lighthouses.
- Explore the dilemma faced by Hamish's father when he is deciding whether or not to use electricity. You could use a thought corridor (see p. 66) with reasons for and against using electricity.

Explore the science that is directly linked to the story:

- Simple series electrical circuits and their basic parts, including cells, wires, bulbs, switches and buzzers.
- A complete circuit is needed for electricity to flow.
- An open switch breaks a circuit

Explore the science that is indirectly linked to the story:

- Many appliances run on electricity.
- Electrical conductors allow electricity to pass through them and insulators do not.
- How to draw circuit diagrams using electrical symbols.
- How to make bulbs brighter or dimmer.

Try these science activities:

ELECTRICITY – SERIES CIRCUITS

1. The discovery lesson

Give out bulbs, batteries and wires and let the children discover how to make the bulb light. If they work it out themselves, they will remember it more clearly than if you show them. It may take ten minutes for the first team to work it out but then they will share what they have learned and there will be a real buzz in the room.

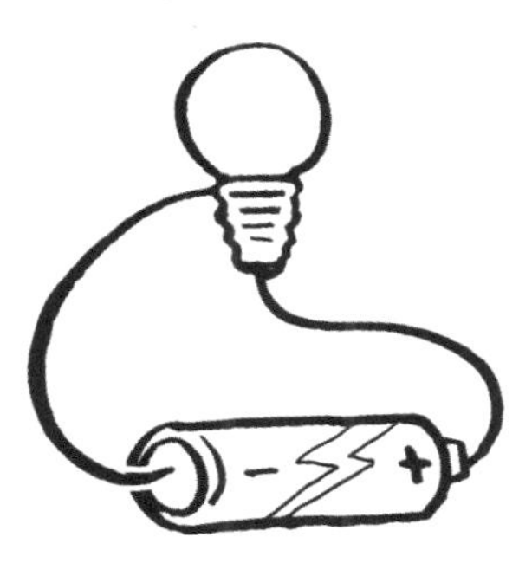

Observe a bulb with a hand lens to see how the wire inside it runs from the side to the filament and back to the base, so the connecting wires must touch the side and base of the bulb.

Remember to explain to the children that the batteries they are using in class are at a safe voltage (1.5 V) whereas the mains wires carry electricity at a dangerously high voltage (240 V) which could kill them if they touched a bare wire.

2. Circuit diagrams

Introduce circuit diagrams. Predict which bulbs will light in a particular circuit and then test each circuit to see what happens. Practise drawing circuit diagrams of the new circuits the children have made.

3. Conductors and insulators

Test and sort materials to find conductors that Hamish could use to fix the broken wire. You'll need to make a circuit with a gap and put the materials across the gap to see if they conduct electricity – by completing the circuit and making the bulb light.

4. Switches

Observe pre-made switches in a variety of circuits to see how many bulbs go off. Make a switch following the teacher's instructions. You could create two terminals with drawing pins and use a paper clip to swing into place and join up the circuit between the terminals.

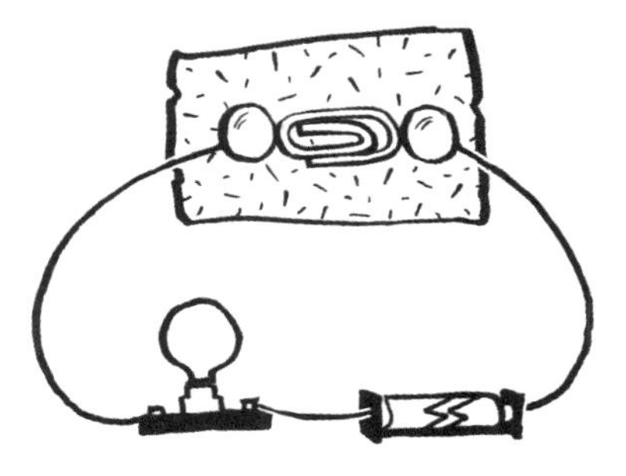

Design your own switch. Look at the design of switches we use in everyday life. You could make pressure switches where you press one terminal down onto another to close the switch. You could make switches to sound an alarm when a vessel is full by making one terminal that can float up on the surface of rising water to close the switch. You could even make pneumatic switches by using two syringes connected by a tube filled with air so that when one syringe is emptied into the tube the air pushes out the plunger on the other end of the tube and this closes the switch.

Continued overleaf

ELECTRICITY – SERIES CIRCUITS *continued*

5. Brighter or dimmer

Investigate ways to make the bulb brighter or dimmer – you could link this to the three lighthouses and how much power the generator would need to power three bulbs in one loop. Try putting lots of bulbs in one loop in a series circuit. Try adding extra cells to brighten a dim bulb. Try adding a very long length of wire into the loop to see how the resistance of the wire affects the brightness.

6. Why do we need electricity?

Research appliances and methods of transport that use electricity. Find out where our electricity is made and whether this is sustainable and clean. Find out what life was like before electricity and how it revolutionised life in the home in the 1950s. Use this research in your non-fiction writing.

7. Discovering parallel circuits

The first version of this story works well with younger children (6–9 years), but the story can be extended for older children using the alternative ending in order to teach about parallel circuits. You might try the following activity in class to encourage the children to discover parallel circuits for themselves:

Give the children one battery, some wires and several bulbs. Ask them to find a way to make more than one bulb light brightly. There are various ways to wire up a parallel circuit; the children will enjoy discovering them for themselves. In our experience you can get at least three bright (2.5v) bulbs in parallel from one 3 volt battery (2 x 1.5 volt AA cells).

Explore other curriculum areas that link to the story:

Reading

Read other stories about islands or lighthouses. With younger children you could use the Katie Morag island stories by Mairi Hedderwick (Red Fox, London, 2010) or *The Lighthouse Keeper's Lunch* by Ronda and David Armitage (Scholastic, London, 2007).

Fiction

Box up and write the story, or innovate by changing the characters and setting, or the solution (fix the circuit another way). You might also recycle the plot using other technologies, e.g. burglar alarms, street lights or electricity supply to a group of new houses.

Non-fiction

Instructions: Write a page from Hamish's book on how to build a circuit.

Instructions: In role-play, direct the workmen how to connect up the lighthouses on the island.

Explanation: Explain why electricity is important in everyday life.

Recount: Write a letter from Hamish to his brothers on the mainland telling them about what happened.

Persuasion: Write a letter or an article in Island News to encourage other lighthouse keepers to use electricity to power their lighthouses.

Geography

Compare life on an island to life on the mainland.

Study the geographical features of an island.

Interpret and draw maps of islands.

History

Research the life of Grace Darling, who was a lighthouse keeper's daughter.

Design/Technology

Make a model island with lighthouses that can be wired up with all the lighthouses in one loop like on Hamish's island.

Topic 5.8 SOUND

Story: The Bat Who Learned to Click

Topic: Sound, sounds travelling through different media, echoes, sound absorption

Plot type: Quest

Genre: Folktale

This is a story about a bat who is ostracised from his colony because he has a terrible cold. While he is away from the colony, he realises that he can use the echoes of his own coughs to locate flies to eat. He returns to the roost with this new skill and teaches all the other bats to hunt in this way.

The science content in the story includes the idea that sounds the bat makes get reflected by the hard body of the fly. It also makes reference to the idea that when the sound has to travel further it takes longer to arrive and that the soft fur of the bat absorbs the sound. It also mentions that the bat's ears are large, which gathers more sound. Although this story suggests a way for echolocation to evolve, it is a work of fiction, belonging to the 'how things came to be' genre, and does not purport to be an accurate scientific account of bat evolution. Like all the stories in this book, this story provides a fictional backdrop to give context to science. You may want to discuss the theory of evolution too at this point so the children can see that there are no firm answers in this area of science and that any description of how a species came to exist is based on fossil evidence rather than eyewitness accounts and is, therefore, open to interpretation.

The Bat Who Learned to Click

Long ago, in the time of the ancients, there lived a colony of cave-dwelling creatures with long fingers and terrible eyesight. They looked a little like moles or shrews and lived in the darkness of the caves, eating the insects – the crawlers and creepers that moved slowly along the ground. But they yearned for more. They longed to eat the tasty flies that flew through the air. They stretched out their fingers to catch the flies but they just couldn't reach. Some of the creatures were born with longer fingers but they still couldn't catch the flies. Time passed and some of those long fingered creatures were born with webbing between their fingers and their hands became wings. They could fly!

They were bats! But they still couldn't catch the flies because they had such poor eyesight.

And then, one day, one bat got a terrible cold. He sneezed and he sneezed and he couldn't stop. The other bats in the roost shuffled away from him and covered their snouts with their wings. So he took himself off, to be by himself and to die.

He curled up by the side of a lake and went to sleep. A few days later, the bat found, to his surprise, that he hadn't died and his sneezing had stopped! He felt better but he was terribly hungry. And right there in front of him were flies, swooping over the lake, looking so delicious but so out of reach. He sighed a disappointed sigh and the breath caught in his dry throat. He coughed. And as he coughed a very fat fly skittered by. And do you know what he heard? A tiny echo of his cough just a half a blink later.

This was no ordinary bat. This bat was a thinker and he thought about the tiny echo and he had an idea. He sat up a little and waited for another fat fly to skitter by ... And then he coughed. And once again the sound of the cough bounced off the fly and came back as a tiny echo.

He tried again, this time with a fly a little further away. This time the squeaky cough took two half-blinks to bounce back. The little bat's thoughts were racing. He pricked up his big ears, all the better to hear the tiny cough echoes, and took to the sky. He flew silently. His soft fur hardly made a ripple in the air.

He spotted a fly and turned to chase it. This time he didn't use his short-sighted eyes to follow the fly. Instead he coughed and he clicked and he listened for the echo and followed it. Faster and faster he chased until the cough echoes came back in less than half a blink and then he opened his mouth and – gulp! His first fat fly! Oh my! ... They tasted even better than they looked.

When he had eaten his fill, he flew back to the roost and told everyone about his new discovery. The other bats in the roost uncovered their snouts and shuffled a little closer to learn how to cough and to click. From that day on, all of the colony listened for tiny echoes of their clicks to find the fattest flies for their dinner.

And still, to this day, many bats hunt by listening to the echoes of their clicks. It all started long ago with that one little bat.

Top tips for telling

This kind of story needs to have an atmosphere of magic and wonder. Establish this at the start in your pace and tone of voice when, as the bats stretch out their fingers, they get longer and longer and then turn into wings. This establishes the 'how things came to be' genre.

After that, the main thing is the drama of the little bat, who thinks he is going to die, then recovers and discovers a way to catch a big fat fly. Keep the wonder in the story all the way through, and evoke celebration at the end when all the bats learn to click.

Ways to work with the story

Teach the story using HMSS

Deepen the story with activities such as these:

- Re-enact the bat experimenting with the coughing and clicking, and the sound bouncing off the fly.
- Role-play the bat explaining how to use echolocation to the other bats in the roost.
- Role-play the bat trying to persuade the older members in the roost to try out this new skill, despite their reservations.
- Role-play being a bat expert presenting a report on echolocating bats.

Explore the science that is directly linked to the story:

- Sounds get fainter and take longer as the distance from the sound source increases.
- Sound is reflected by hard surfaces.
- Sounds can vary in volume.
- Some animals use sound to echolocate.

Explore the science that is indirectly linked to the story:

- Sounds are made when something vibrates.
- Sound vibrations travel though a medium to the ear.
- The structure of the ear and how to protect our ears.
- Sounds can vary in pitch.
- Natural selection and the theory of evolution.

Try these science activities:

SOUND

1. Vibrations – the discovery lesson

Observe vibrations :

- touch the prongs of a vibrating tuning fork onto the surface of a bowl of water;
- dangle a bead on a string next to a vibrating tuning fork;
- feel vibrations in your voicebox by touching your throat with your fingertips;
- put rice on a drum and watch it leap as you play the drum;
- watch videos of cornflour monsters – there are lots on the internet (cornflour and water are mixed and placed on a metal tray on a speaker playing loud music, and the mixture seems to dance);
- stretch a balloon like a drum skin over the end of a cardboard tube and fix it into place. Speak in to the open end and you can feel the vibrations on the balloon.

2. Sound travelling

Sound waves are compression waves where air molecules are compressed in waves by vibrating objects. Demonstrate this motion with a Slinky toy. Stretch out the Slinky and hold one end while someone else holds the other. Flick your end inwards, towards the other end. Watch the wave travel up the Slinky.

Use tuning forks to make a sound that will travel through different materials. Test whether sounds are carried most effectively by hard or soft materials by touching the base of the tuning fork onto one end of an object and listening to see if you can hear the sound at the other end of the object. You could use a soft pillow, for example, and compare how well the sound travels through it compared with a hard wooden table.

3. Volume

Investigate volume – what makes a sound louder or quieter? Make paper-cup squeakers of different sizes. You can make a paper-cup squeaker by anchoring a string in the base of a cup so that the string hangs down inside the cup and beyond the rim when the cup is upended. Wet the string and drag finger and thumb down the string to make it squeak. The cup amplifies the noise.

4. Pitch

Investigate pitch – what makes a sound higher or lower in pitch? Make reeds/squeakers of different lengths from paper straws.

Make elastic-band guitars of different tightness, strike chime bars/triangles/spoons of different lengths. Strike bottles of different sizes or filled to different heights with water. Talk about what you observe.

Continued overleaf

SOUND *continued*

5. Echolocation

Research other animals that echolocate to sense what's around them, such as dolphins.

6. The ear

Learn about the structure of the ear and how it enables us to hear different sounds. Learn about how the ear is sensitive and how to prevent damage.

Explore other curriculum areas that link to the story:

Reading

Read other similar stories in such collections as *Just So Stories* by Rudyard Kipling (Oxford University Press, Oxford, 2009) or *How the Whale Became: and Other Stories* by Ted Hughes (Faber & Faber, London, 2011).

Fiction

Box up and write the story, or innovate using different animals that echolocate (for example, whales and dolphins).

Non-fiction

Instructions: Write instructions with diagrams on how to catch a fly using echolocation.

Explanation: Write an explanation of how bats (or other animals) use echolocation to catch their prey.

Persuasion: Be a bat and persuade Granny Bat to use echolocation.

Information report: Give a presentation on the features of a fictional echolocating dragon.

Information report: Write a report entitled 'All about Bats'.

Life Skills

Find out about human echolocation by people with visual impairment, e.g. Daniel Kish or Ben Underwood.

Music

Find out about Evelyn Glennie, a deaf musician who can play by sensing the pitch of instruments by using her whole body to detect their different vibrations. She feels certain vibrations in her cheekbones and others in her chest.

History

Research the invention of sonar or radar.

Research Alexander Graham Bell and the invention of the telephone.

Topic 5.9 SPACE

Story: Apollo 13

Topic: Space travel, orbits, gravity, the moon
Plot type: Voyage and return
Genre: Historical fiction

This is a simplified telling of the voyage of the fated Apollo 13, which was supposed to land on the moon. It suffered damage to the oxygen tanks and was forced to abort the planned landing. The support team back at Ground Control had to work out how to get the crew safely home by working together and finding solutions to all the problems, using only the resources available on the spacecraft.

We include this story partly to present science and scientists in a heroic light, to encourage children to think of science as a possible career. The story also teaches that we should never stop trying. It's a great story with real drama and it can be illustrated with real photographs, freely available on the internet, along with the audio clips of *Houston, We Have a Problem!* and the film *Apollo 13* directed by Ron Howard (1995). The film was rated PG, so you may need to seek permission to show it to your class.

The science content in the story includes gravity and space travel.

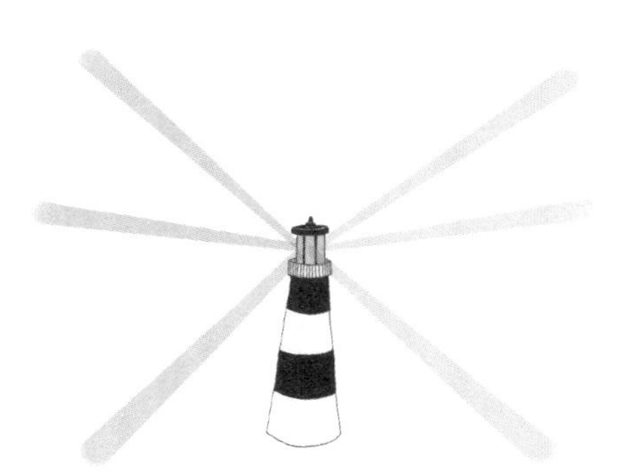

Apollo 13

Once ... there was a space mission. It was called Apollo 13, the third manned spacecraft to go to the moon.

It was 11 April 1970.

The astronauts' job was to explore the hills around a huge moon crater to see if they could collect any pieces of the meteor that had made the crater.

Three astronauts – Jim, Jack and Fred – were aboard and excited to be on their way. They waved to their families and strapped themselves into the spacecraft. The clock said 13:13. It was time to go.

10 – 9 – 8 – 7 – 6 – 5 – 4 – 3 – 2 – 1 – lift off!

The rockets blasted and the spacecraft rattled its way up into the sky. It wasn't long before it was just a speck to the crowd cheering below.

Inside, the vibrations stopped and the crew felt themselves lifting off their seats as they became weightless. All was well. The launch had been successful. Jack pulled himself along to a porthole for his first glimpse of the Earth. Jim turned on the radio. He chatted to Houston Ground Control as he flicked the switches and read the readings on the dials all around him.

Two days passed. The astronauts got used to being weightless and learning to eat and sleep while floating in the air in the cramped space of the command module. They were all excited about landing on the surface of the moon in a few days' time. Then, on the third day, they were going through the usual routines, as instructed by Ground Control, and Jack flicked the switch to stir the oxygen tanks –

BANG

There was a huge explosion. The rocket was knocked off course. Fred grabbed at the handles on the wall and pulled himself over to the porthole and looked out.

'Houston, we have a problem,' said Jim on the radio.

'Check the gauges!' cried Fred. 'I think we may have hit something. Maybe something exploded.'

Jim checked the gauges and the levels of oxygen were falling fast in tank number two. And then it was reading empty. They turned their attention to tank number one and the levels were falling slowly in that tank too. The explosion must have damaged it. They would soon run out of oxygen altogether at this rate.

In fact, one of the huge oxygen tanks had been ripped open. There was a huge hole in the tank. The panel had blown right off.

'Jim, you need to take the men into the lunar module,' ordered Ground Control. 'Now!'

The three men grabbed what they could from the command module and squeezed into the lunar module and shut the hatch. It was cramped in there. It was the part of the spacecraft which was designed to actually land on the moon. It wasn't meant to be a living space. There the three men waited ... in the cold, dark silence of space.

At Ground Control everyone was busy. Mathematicians calculated how long the men could survive ... the news wasn't good. Engineers looked at the damage and searched for a way to repair the hole ... the news wasn't good. Physicists looked for a quick way home ... the news wasn't good.

And the men on Apollo 13 waited. They switched off the lights to conserve the battery power. They sat still and controlled their breathing to save their oxygen. They even turned off their radio to save power. And there they waited ... in the cold, dark silence of space, not knowing how long they had to live.

Back at Ground Control, the Director refused to give up. He called the mathematicians, he called the engineers and he called the physicists.

'Find a way to get our boys home,' he said, banging his fist on the table. They looked at one another. They thought it was hopeless; there was just so little time before the oxygen and the fuel ran out. 'Find a way to get our boys home,' the Director insisted.

They began to think. They had to get the spacecraft back on course ... and re-enter the atmosphere ... and land ... and so much of the Apollo 13 was broken.

They laid out on a huge table every working thing that was on board the craft. It wasn't much.

'Think!' said the Director. 'Their life is in your hands.'

TOOL
PULL
OXYGEN
LEVELS
LOW
DANGER!

PHOTOCOPIABLE FOR USE IN SCHOOL

'Hmmm, this might work,' said a mathematician.

'Hmmm, this might work,' said an engineer.

'Hmmm, this might work,' said a physicist

Little by little they came up with ideas: things from the lunar module could be adapted to fit in the command module and allow them to breathe. Carefully planned blasts from the engines could put them back on the course home. There was hope.

When Jim put the radio back on, all kinds of instructions came through from Houston. All three astronauts leapt into action. They were glad to be busy. They went back into the command module, where they had to reconstruct a gadget to remove the carbon dioxide from the air, using only the bits and pieces they had on board.

Then they had to get the spacecraft into the right position without wasting any fuel on turning the spacecraft around. They waited until the spacecraft had reached the pull of the moon's gravity. They let the spacecraft orbit around the far side of the moon and back around until the spacecraft was travelling towards the Earth again.

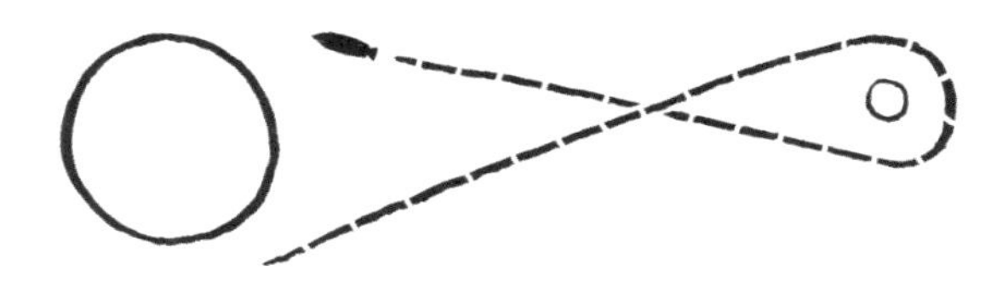

Then they made careful rocket blasts so the rocket headed out of its orbit around the moon and back towards Earth. Finally, as they approached the Earth they released the lunar module into space. The three men watched through the window as their little lifeboat floated away.

Minutes later, the radio went silent as they re-entered the Earth's atmosphere. They were alone again. The spacecraft began to shudder and judder and it got hotter and hotter and hotter. The crew held their breath and hoped. The cabin got hotter and hotter and there was a terrible sound as if wind was blowing inside the cabin. Were they all about to die? Then, suddenly, it all stopped and for a few moments there was quiet ... just the sounds of their own beating hearts ... and then a splash.

They had landed in the sea!

James tried the radio again. 'Hello, hello ... Houston?'

And the room at Ground Control erupted with cheers.

They were home.

Top tips for telling

This story can be told as an action drama. Evoke first the excitement and optimism of the launch and then the dread as the oxygen tanks are broken. After that, there's the thoughtfulness and determination of the Director not to give up and let his men die.

When the spacecraft re-enters the atmosphere, make it clear that the men know they might easily die. The rocket could easily explode. Then emphasise their relief as they splash down in the sea.

You might like to use photographs of the different parts of the spacecraft, or even a model, to help you tell the story.

We have not developed individual characters much; the main thing is to evoke the drama of each situation and the determination of the Houston scientists to find a solution.

Ways to work with the story

Teach the story using HMSS

Deepen the story with activities such as these:

- Explore the movement of the sun, moon and Earth by acting out the rotations and orbits in movement or dance, or using balls of various sizes.
- Re-enact a conversation between an astronaut and his child in which the astronaut explains how the rocket works and what it's like in space.
- Re-enact an interview, before the flight, on a TV show where an astronaut explains the purpose of the flight and what they hope to learn.
- Role-play conversations between the astronauts as they worry about what will happen when they fall back to Earth.
- Role-play the TV news reporting on the events as they happen in the story: first the launch and explaining how the rocket works and how it gets away from gravity; then the report on the explosion and, finally, as the spacecraft returns to Earth, explaining why gravity pulls the craft down towards Earth and why it gets so hot.
- Role-play a press conference once the astronauts are safe.
- Write press articles and features about the events.
- Draw a labelled picture of the spacecraft and explain what all the parts are for.

Explore the science that is directly linked to the story:

- Unsupported objects fall towards the Earth because of the force of gravity.
- The moon orbits the Earth because it is held in orbit by the gravitational pull of the Earth.

Explore the science that is indirectly linked to the story:

- how the planets move in the solar system;
- the Earth rotates to give us day and night and this rotation gives rise to the apparent movement of the sun across the sky (see activities on p. 154).

Try these science activities:

SPACE

1. The sun, the moon and the Earth

Study their relative sizes and distances from each other. Demonstrate this in as many ways as you can: videos, computer models, and balls of various sizes laid out on the school field.

2. Gravity and the planets

Find out about the relationship between gravity and the mass of the planets. Find out how it holds moving bodies in orbit. Research the names of the planets and their relative sizes, masses and gravity. Make Top Trumps style cards to help the children become familiar with basic facts about the planets, such as size, mass, gravity compared with Earth's gravity, distance from the sun, day length and surface temperature. Explain how the gravity of the sun keeps all the planets in orbit.

3. Night and day

Make sundials to demonstrate how the rotation of the Earth causes day and night and the changing position of the sun in the sky. See how the position of the shadow moves as the Earth turns.

4. Orbits and rotations

Learn how the sun and moon and Earth move in relation to one another to cause the phases of the moon and the seasons. This can be done with models such as an orrery set or with videos from the internet – the BBC Science Clips website is a good example. Children also find it helpful to move like the Earth or moon to fully understand how the Earth and moon are spinning (rotating) and moving in a circle around the sun (orbiting) at the same time.

5. Constellations

Find out how to find the Pole Star and well-known constellations. Invite a portable planetarium into school or organise a star-gazing evening. Look at maps of the constellations and encourage the children to look at the night sky at home.

6. Innovations in space exploration

Use the internet to find out about the development of Curiosity, the Mars rover. Learn how it was designed to use very little energy and can be programmed to explore certain areas and carry out tasks such as collecting samples. Find out how it has been designed to manoeuvre on the surface of Mars. Find out how it communicates using the Mars Odyssey (in orbit around Mars) to relay information back to Earth and to receive its instructions.

Explore other curriculum areas that link to the story:

Reading

Read other stories from the history of space travel.

Read *The Sea of Tranquility* by Mark Haddon (HarperCollins, London, 2008).

Fiction

Box up and write the story, or innovate details about invented problems faced by the astronauts and how they were solved by Ground Control.

Non-fiction

Information report: Write fact files about the Earth and the moon, the planets, spaceships or aliens.

Recount: Write a diary entry for an astronaut.

Recount: Write a newspaper report of the launch, the problem or the homecoming.

Discussion: Would you like to be an astronaut? Discuss the pros and cons of being an astronaut.

Discussion: Is anyone out there? Have the class debate the case for the existence of aliens.

Discussion: You could have a fictional discussion such as 'Should Daleks Be Allowed to Live on Earth?' – from *Talk for Writing across the Curriculum: How to Teach Non-fiction Writing 5–12 Years* by Pie Corbett and Julia Strong (Open University Press, Maidenhead, 2011).

Explanation: Write a presentation for a news programme explaining what went wrong and how the problems were resolved.

Art

Look at examples of science fiction artwork. Use shading to recreate the style of science fiction graphic novels. Sketch still-life compositions of interesting metal objects in this style.

Design alien spaceships and make sketches of alien life forms.

History

Find out about Neil Armstrong and the first landing on the moon.

Discover the importance of the constellations to ancient civilisations or to the explorers.

Design/Technology

Design and make a model of the Mars rover with moving wheels using a cam and axle. Invent your own equipment to survey the surface of Mars.

Design, make and taste your own 'space food' out of freeze-dried ingredients.

Music

Listen to sections of *War of the Worlds* by Jeff Wayne (1978), then create a mime or dance to go with the music, retelling the story, or create artwork or a play based on the story.

Topic 5.10 ELECTRICITY – PARALLEL CIRCUITS

Story: The Rocket

Topic: Electricity – parallel circuits
Plot type: Voyage and return
Genre: General fiction

This is a story about an old professor whose ambition is to go into space. He builds his own rocket, but then faces a crisis in space when things need rewiring. The story is left on a cliffhanger and provides a problem that can be solved using parallel circuits. The children can then decide how they wish to end the story.

The science content in the story includes making circuits and putting more than one component into a series circuit so that a bulb becomes dimmed. It requires the teaching of parallel circuits in order to solve the problem and so save the professor's life.

The Rocket

There was once a professor of the stars. He knew everything there was to know about the moon and the stars and the planets. All his life he had wanted to go and see them for himself, but his wife wouldn't hear of such nonsense.

'Alfred my darling,' she'd say, 'you can't even wire a plug. How on earth could you make it to space?'

Alfred adored Agnes and would never do anything to upset her, so he bought a telescope instead and spent hours watching the stars from the safety of his back garden in his old deckchair, while Agnes fixed the leaky pipes and put up the curtain rails and mended the squeaky floorboards.

Then, quite suddenly, Agnes died. And Alfred found himself alone in the house. There was no one to talk to, no one to drink tea with and only one pair of slippers by the bed. It was a strange and lonely time. Alfred spent a lot of time stargazing. He wondered about selling Agnes's shiny red car to pay for a new telescope, to cheer himself up, when suddenly it occurred to him ... Agnes was gone ... he was free to go to space! All he needed was a rocket. He would turn her old car into a rocket!

He enlisted the help of some other retired professors from his old university days. They peered over their half-moon glasses and drew up plans, glad to be doing something again. While Alfred brewed endless cups of tea, his friends tinkered and engineered, stroked their beards and got on with converting the little red car into a rocket. Alfred went to the hardware shop to get all the things they needed and stopped on the way home at the library for *The Electrician's Handbook* – because his wife had been right: he might know everything about space but he really couldn't wire a plug. He ordered the fuel on the internet and his friends carried on, working day after day until, at last, it was finished. There, in the yard, was a bright red shiny rocket.

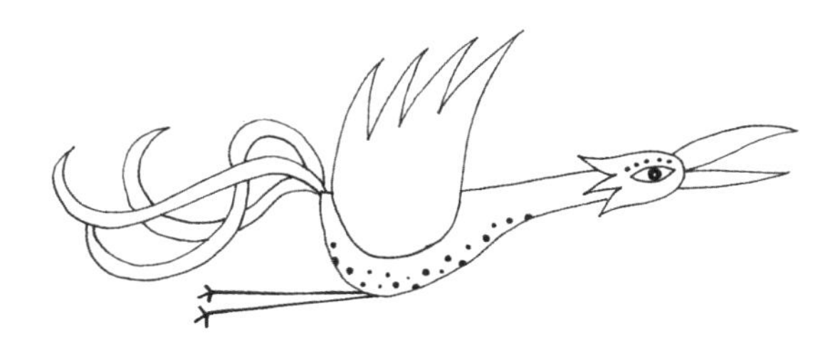

Alfred packed his bags with food for a week, a telescope, a map and *The Electrician's Handbook* (just in case). Then, with his retired professor friends waving sagely and calling out instructions, he blasted off.

It was a bumpy ride through the atmosphere until the professor felt himself lifting out of his seat – weightless at last. The view was tremendous but the temperature was dropping rapidly, so the professor put on the heaters and rubbed his hands as he got out his telescope. He gazed in wonder at the stars around him. 'See, Agnes, I did it ... I finally got into space!' he muttered as he peered through the lens.

The rocket drifted slowly moonwards and the professor turned out the lights to see the stars more clearly. Oh the glory of the heavens! The professor stared and stared and stared. The rocket drifted slowly on, pulled by the moon into orbit around it. He travelled right around the moon and back towards the Earth. The professor had never seen such beautiful stars.

Time passed and the professor was nearing the end of his journey. It would soon be time to land. He sighed and tore himself away from his stargazing and looked about for his map. In the dim light, the professor could barely see the map, so he reached for the light switch. The lights flickered and went off. At that same moment the heater went off too. Alfred was left in the dark and the cold.

The professor frowned at the heater in the dim starlight and thought for a moment. Then he realised that the batteries had both run out. 'No matter,' he muttered as he pulled out the wiring from the dashboard, 'I have spares,' and he opened up the glove compartment.

Inside, there was chocolate, spare socks and one spare battery.

'One? But I need two!' he gasped. He rummaged around under the chocolate and the socks but there was only one battery. Alfred needed two – one for the lights and one for the heater. He needed lights to read the map on re-entry or he'd not be able to land in the sea. If he hit land he would die for sure. But if he didn't get some heat soon he'd freeze to death anyway. Which should he choose, the map or the heater? His professor friends weren't here to help him now!

He laid all the things out on the floor: battery, wires, bulb and heater. He could either join the battery to wire to heater to wire and back to battery all in one loop, or battery to wire to light to wire and back to battery all in one loop.

Which loop should he choose? Crash land or freeze on the way down?

'Oh Agnes!' moaned the professor as he rubbed his temples and stroked his beard. He flicked anxiously through *The Electrician's Handbook* but the diagrams just spun before his eyes.

What happens next? What did the Professor do? What would you do if you were him? Which circuit would you choose? How would you solve the problem?

Stop the story here!

Your students can choose whether the professor resolves his dilemma in time to land safely – or works it out too late to save himself. To solve the problem, the children will need to know about parallel circuits and how you can wire two bulbs in parallel so that neither will dim. We have suggested activities to teach this at the end of this chapter. Once you have taught the science you can revisit the story and the class can create new endings based on what they have learned.

If a student suggests wiring the heater and light together in a single loop, you might respond by saying that the professor tried this but that the bulb was too dim and the heater did not have enough power, and thus lead them to the parallel option.

Top tips for telling

This is a rather quirky fantasy. You may tell the beginning with a little irony to encourage your students to suspend disbelief about the car becoming a rocket. Then evoke the wonder and excitement of space, and the sights the professor sees. The professor's character is key: not very good at practical matters, a loving husband and passionate about space.

When the light and heater go off, be really clear and specific about the problem. You may want to present the dilemma using a battery, some wires, a light bulb and an object representing the heater, or else to draw the situation on the board.

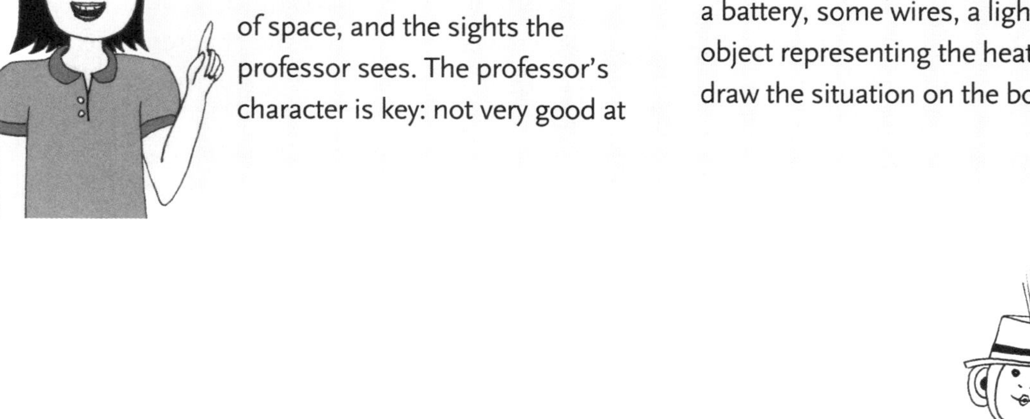

Ways to work with the story

Teach the story using HMSS

Deepen the story with activities such as these:

- Explore the circuit by stepping the path of the electricity from the battery all the way around the loop.
- Re-enact the flight of the rocket, showing the route that the rocket took in a figure of eight around the moon and back around the Earth (see diagram on p. 217).
- Role-play the professor solving the problem of rewiring the circuit to power two components using one battery.
- Role-play an electrician on the radio explaining to the professor how to rewire his circuit.

Explore the science that is directly linked to the story:

- series circuits (see p. 203);
- how to make the bulbs brighter or dimmer;
- parallel circuits.

Explore the science that is indirectly linked to the story:

- electrical resistance;
- the gravitational pull of the Earth and the moon and why the rocket would be pulled into orbit around them;
- the moon landings.

Try these science activities:

ELECTRICITY – PARALLEL CIRCUITS

1. Review basic circuits and switches

If your class hasn't studied electricity recently, make sure you recap all the basics (see the activities on p. 203 and 204) Let them work out how to wire a circuit with no bulb holder so they find out that the wires must touch the side and the base of the bulb because the wire inside the bulb runs from the side to the base of the bulb. Let them use switches to see that a break in the circuit turns everything off because the electricity won't flow. Recap safety with electricity.

2. Circuit diagrams

With a range of different components such as switches, bulbs, motors and buzzers, ask the children to construct the circuits according to circuit diagrams. They can also practise drawing circuit diagrams accurately using all the symbols they have learned. Help them to pay attention to small details such as which way a cell faces in the circuit.

3. Solve the professor's problem

Revisit the story. Give out the equipment that the professor had: one battery, four wires, a bulb and a heater (use a buzzer to represent the heater, since it will sound different at full power and half power). Allow the children to explore different ways to connect up the circuit. If you give them time, at least one group will discover how to wire the bulb and buzzer in parallel and find that both work at full power. Allow those who work it out to help the others. Draw the circuits they have made – parallel circuits can be laid out in more than one way.

4. Fixing problems

In advance, prepare a range of electrical circuits with deliberate errors, e.g. tiny breaks in the wire, opposing cells, bulbs not screwed in, etc. Give the children time to investigate the circuits and work out why they don't work.

5. Brighter and dimmer – altering the current

Provide some circuit diagrams for the children to test out. Include circuits with two bulbs in series, more powerful batteries and parallel circuits. Talk about how to make the bulbs brighter or dimmer. You could also include elements such as a long stretch of wire to show that even wire itself provides resistance in a circuit and long wires will dim the bulb. You could include graphite, which is such a poor conductor that even an 8 cm piece will provide enough resistance to dim a bulb.

6. Measuring brightness

Construct a set of series circuits, first with one bulb, then with two and so on. Demonstrate that it is possible to measure brightness by seeing how many layers of tissue can be placed over the bulb before the light is obscured. It is different for different colours of tissue. If you want to do this quickly, use black tissue! Count how many layers you need to obscure the light when there is one bulb, then two in series, etc. You can graph these data. Remember to always use the same bulb when covering a bulb with tissue, since bulbs all vary slightly in brightness.

Explore other curriculum areas that link to the story:

Reading

Read other books about space travel or the stars. A good non-fiction text is *Space: A Children's Encyclopedia* (Dorling Kindersley, London, 2010).

Fiction

Box up and write the story, or innovate using different problems to be solved in space, or recycle the plot on a different journey, for example driving across a desert or taking a boat to a distant island. You might weave in solar power, electric cars and other electrical technologies.

Non-fiction

Instructions: Write a page for *The Electrician's Handbook* on how to wire a heater and lamp in parallel.

Instructions: Be the other professors and instruct Alfred on how to launch the rocket or how to solve the problem.

Recount: Write the professor's speech or article for the Royal Institution about his journey.

Information report: Write a report about the solar system.

Explanation: Write an interview for a paper in which Alfred explains how he solved the problem.

Explanation: Explain to a friend why a bulb lights up.

Persuasion: Be Agnes persuading Alfred not to go into space.

History

Investigate the history of space travel.

Life Skills

Explore the items that you would take to space if you had built a rocket.

Design/Technology

You can make pop rockets using photographic film pots. These little pots explode and send the base of the pot flying into the air. If you put a couple of centimetres of water into the pot, add an effervescent vitamin tablet and snap on the lid, the bubbles of carbon dioxide formed will be enough to propel the pot upwards several metres. You must do this outside and put the pots lid-downwards on a flat surface for the best trajectory. You can change the amount of tablet and water to investigate which amount propels the rocket highest. You could also design nose cones and fins to help it to fly straight. Once the rocket is set up, stand well back until the rocket has popped. You can order little pots from good educational catalogues if you can't get them from a photo printing outlet.

Topic 5.11 FORCES AND BRIDGES

Story: Bernie's Bridge

Topic: Forces in opposition, strong structures, bridge design absorption
Plot type: Quest
Genre: General fiction

This is a story about a farmer's son who is sent to look after the sheep in an isolated meadow, which can only be accessed by a bridge. When the bridge falls down, Bernie must build another in order to get the sheep home. The story is left, quite literally, on a cliffhanger and the children must invent their own ending to save the sheep. This may be told orally or written down.

The science content in the story centres on bridge design. This opens up to investigating the forces involved in building bridges, and measuring the forces acting on a variety of bridge types that can be modelled in the classroom.

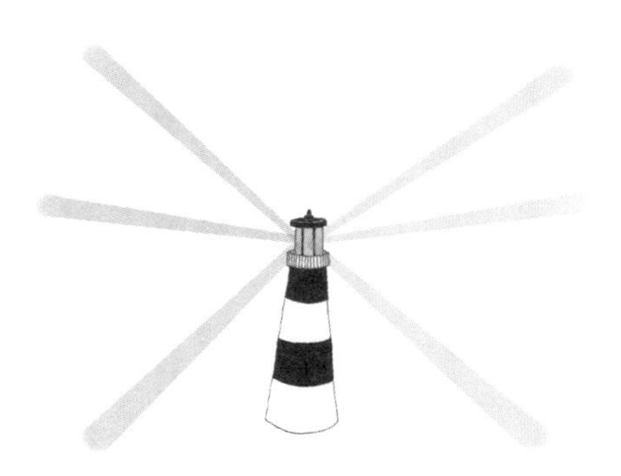

Bernie's Bridge

Once ... there was a farmer with three sons. The youngest of the three was called Bernie. Bernie was as strong as an ox and worked hard, but tended to forget things. Cows escaped when he forgot to close the gates. Haystacks rotted when he forgot to cover them in the rain. The only thing he was any good at was looking after the sheep. He loved the sheep. They were soft and warm and they liked Bernie. The sheep never complained if he made a mistake.

His older brothers were fed up with all Bernie's forgetfulness, so they sent him to tend the sheep for the summer in a distant green field far from the farm.

'Off you go, Bernie,' they said. 'Go and make yourself useful ... on the other side of the valley.'

'He can't go on his own!' their father exclaimed. 'One of you will have to go with him and make sure he doesn't get lost.'

Bernie was about to insist that he'd be fine by himself when his eldest brother, Wilf, rolled his eyes and set down his tools. 'I'll go. We can't have the sheep getting lost or falling down the side of the mountain. We need them to sell at the market.'

Bernie glared at his father and glared at his brother.

'Don't worry,' said Wilf. 'I'll only stay until the morning, then I'll come back to the farm and leave you with your precious sheep.'

So, Bernie and his brother packed a tent and a cooking pot and some tools and set off in the pony and cart, herding the sheep along the road before them. Giddy up! The last part of the journey took them over a bridge across a deep, fast-flowing river to a lush meadow beyond. The meadow was surrounded on three sides by steep mountain slopes and on the fourth side was the river. The only way in and the only way out was the rickety old bridge over the tumbling river below.

As Bernie and Wilf followed the sheep across the rickety bridge on the pony and cart, the bridge creaked and groaned. 'There you go my lovelies!' Bernie said, herding the sheep towards the meadow on the far side.

The brothers unpacked the cart. They took out the tent, but ... Oh no! Some of the lambs were wandering out of the meadow and back over the bridge. Bernie ran after the lambs and shooed them back safely to their mothers in the meadow. The old bridge creaked and groaned with every step.

They got the tent up and Wilf went off to look for some firewood while Bernie unpacked the cooking pot and the food, but ... Oh no! The lambs were wandering away over the bridge again. Bernie leapt up and went after them but the lambs had already reached the other side. As Bernie stepped onto the bridge, it creaked and groaned and cracked, toppled sideways and collapsed, with a splash, into the river. Bernie fell down into the seething water.

The river ran deep and fast at this point, tumbling Bernie over and over, further and further downstream. But Bernie was strong and he managed to claw his way to the edge. He scrambled up the bank, heaving himself up the rocks and the bushes and back into the field. The mother sheep were all bleating in the meadow on one side of the river. On the far side the lambs were bleating back. Bernie didn't know what to do. He couldn't carry the lambs through the water – it was far too deep and fast flowing. And he didn't have much time. The mother sheep were getting more and more upset. Soon, they might jump off the bank to get to their lambs – and drown.

Bernie called for his brother, who came running back holding armfuls of firewood.

'What happened to you?' Wilf laughed when he saw his brother soaking wet and muddy from scrambling up the steep riverbank.

'The bridge – it broke!' explained Bernie crossly. 'And my lambs are all stuck on the other side of the river. We'll have to build a new bridge, and fast.' Bernie looked at the wood in Wilf's arms. 'Where did you find all that wood?'

Some of it was sections of planks and beams. Wilf pointed to a pile of wooden planks in the corner of the meadow from a barn that had blown down in a storm. Together, Bernie and Wilf ran back to the pile to see what they could salvage. Maybe they could build a new bridge with the wood!

First, they took two long planks, long enough to reach between the banks of the river, and carefully placed them on the grass. Next, they sawed one hundred short planks and nailed them across the longer ones, like a train track. After that, they practised driving the cart along the wooden track. The wood seemed strong and it neither creaked nor groaned. Bernie was delighted and he and Wilf hefted it with their great strong arms to the edge of the rushing river. Bernie swam out to a large rock in the centre and climbed to the top. Holding his arms right over his head, he was able to guide the track as Wilf pushed it slowly out over the river. It reached the other side and looked firm.

Dripping wet and muddy, Bernie clambered back up the river bank.

'Off you go then,' said Wilf. 'Your lambs are waiting. Go and fetch them.'

'No, not yet. Not before I've tested the bridge. I need to know it will hold the weight of the lambs,' replied Bernie.

Wilf raised his eyebrows just a fraction. Maybe Bernie wasn't so daft after all.

Leaving the pony in the field, Bernie began to push the cart out across the bridge. As he reached the centre, the weight of the cart pulled the bridge down. It began to bend and bow in the middle, as there was nothing to support it. The cart slipped down further and the bridge bowed even more, until Bernie could see it was going to snap. Just before it broke, he jumped out of the way of the falling cart and landed in the river. Once again, the rushing water tumbled Bernie over and over, further and further downstream. Once again, he struggled to the edge. Wilf was waiting for him and helped to haul him up the steep bank.

It was getting late. The lambs were cold and hungry and crying for their mothers. Bernie was desperate. 'Oh no!' he wailed. 'Wilf, what are we going to do now?' There was plenty of wood left in the pile and some rope but they needed a better bridge design.

What happens next? What would you do next if you were Bernie? What kind of bridge would you build across the river? The river is too fast to carry the sheep across and there are no bridges for miles in either direction. All you have are wooden planks, a saw, a hammer and nails, a pile of stones and a long coil of strong rope. In some parts of the field there are tall trees by the river bank. What would you do?

Stop the story here!

Once you have studied bridges with your class you can set the children a challenge in the classroom to build a bridge that will span a given gap and hold a certain weight. They could make strengthened beam bridges, or bridges with a central pillar to hold the weight, or some kind of simple suspension bridge. You can allow a choice of building materials or limit the materials to paper and find ways to make it stronger. (See the activities below.)

After plenty of talk time to practise explaining their ideas and bridge designs, the children will be able to finish the story with Bernie designing a successful bridge to bring the sheep home. Their stories may involve Bernie and Wilf making a few attempts to build bridges before they are successful. Encourage the children to make the sheep sound increasingly anxious as time passes.

Top tips for telling

First introduce Bernie. His qualities are strength and foolishness and a deep love for his sheep. When he drives his cart over the river describe it quite precisely: it is wide and fast flowing. When the lambs are separated from their mothers, make sure you describe where the sheep and lambs are and then evoke their panic. Finally, describe the first bridge precisely so that the whole scene can be clearly visualised by your audience: river banks, river and bridge made from two long planks and 100 short ones nailed across.

Ways to work with the story

Once groups have created their own story ending they can develop their story using HMSS.

Deepen the story with activities such as these:

- Explore the feel of forces that work in opposite directions by playing tug of war.
- Re-enact the part of the story where Bernie builds a basic beam bridge and it bends in the middle under the weight.
- Role-play Bernie explaining why his first bridge failed.
- Role-play Bernie talking to a bridge expert who can advise him on a suspension bridge design.
- Make models and act out the story with model people and sheep. You could even make a film of it.

Explore the science that is directly linked to the story:

- the force of gravity acting between the Earth and a falling object;
- the forces acting on structures like bridges;
- the role played by Isambard Kingdom Brunel in the history of science;
- how different forces can act in opposite directions at the same time (for example, gravity pulls down on the beam of a suspension bridge while the wires pull the beam upwards).

Explore the science that is indirectly linked to the story:

- Other kinds of opposing forces like gravity pulling downward on a falling object while air resistance pushes upwards on the same object, slowing the speed of fall.

Try these science activities:

FORCES AND BRIDGES

1. Brunel and bridge design

Research Isambard Kingdom Brunel and the bridges he designed. Learn about different bridge designs.

2. Measure forces

Investigate the forces acting on bridges. Make some paper bridges by rolling paper into tubes to form a beam and four supporting legs. You could make a simple beam bridge, supported by two legs at either end, then make others of the same basic design with extra legs as supports along the length of the beam. Measure the amount of mass each bridge can support in the centre of the beam before the paper crumples and the mass falls. Which can hold the most? Relate the mass held to the one that withstands the greatest force.

3. Strengthen paper

Investigate ways to make paper stronger, e.g. folding, layering and making columns. Test them by measuring the amount of force each can withstand.

4. Suspension bridges

Make simple suspension bridges using straws anchored to tabletops to form the side pillars and beam. Add a string leading up and over one pillar, looping around the centre of the beam and up and over the other pillar. Stick the ends of the string down to the table to anchor it. Test how much mass the beam will support. Then remove string and test again. You should find that it supports a much greater mass with the string providing support from above in the centre. There are more detailed instructions about how to make suspension bridges on the internet (for example: www.scientificamerican.com/article/suspension-science-how-do-bridge-designs-compare/).

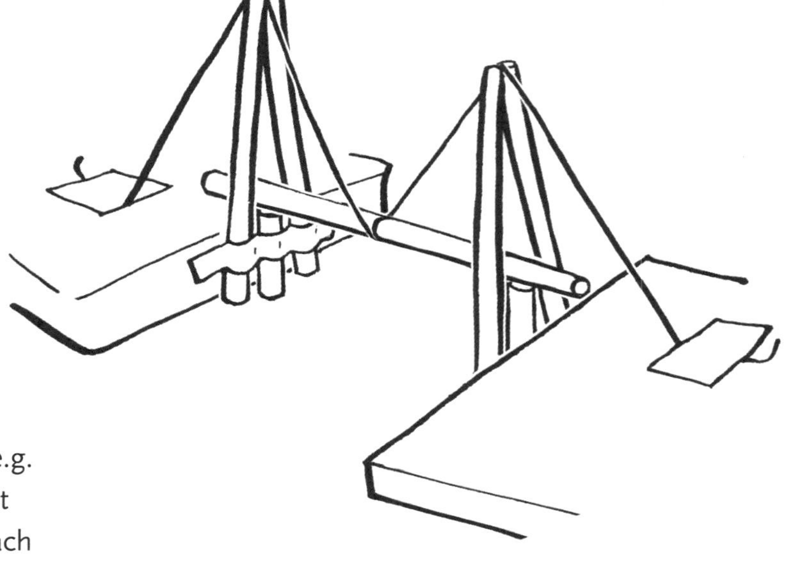

5. Forces on falling objects

To investigate the forces acting on falling objects, make spinners. Note how the spinners with larger wings fall more slowly. Show how Galileo tested his theory that objects of similar shape but different mass fall at the same rate. It's the shape not the mass that makes the difference.

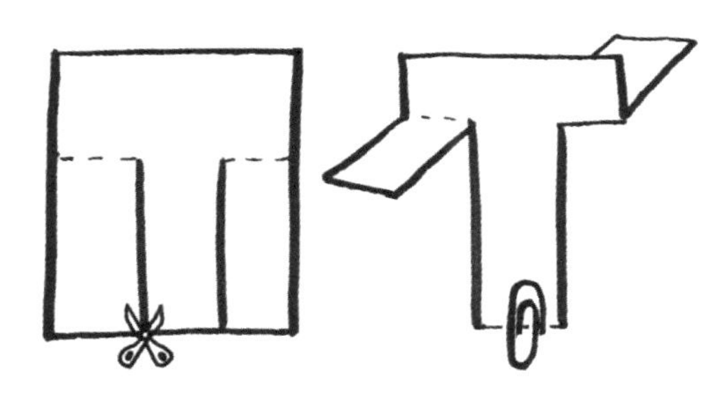

6. Forces in opposition

Investigate the drag forces acting on a car – friction and air resistance. You could relate this to the Bloodhound project which is attempting to break land speed records. You could also investigate the forces acting on a boat. Look at the way the drag forces act in opposition to the wind or motor pushing the boat along.

Explore other curriculum areas that link to the story:

Reading

Read other traditional tales that use the rule of three, such as *Goldilocks and the Three Bears*. There are three options but only the last one is just right. You can find a nice version of *Goldilocks and the Three Bears* in *The Nursery Storybook* by Georgie Adams (Orion, London, 1996). Encourage the children to think about a second and third bridge design for Bernie to build so that they can write endings where the third bridge design is successful.

Fiction

Box up and write the story with an invented ending based on what the children have learned in science lessons. Include three bridge designs if you can, where the last bridge is successful.

Non-fiction

Instructions: Write instructions on how to build a sturdy bridge (suspension, clapper, arched, etc.).

Information report: Write a report about famous bridges and the engineers who built them.

Persuasion: Advertise a bridge design for a bridge company in a TV or magazine advert.

Explanation: Write an explanation, with diagrams, of why Bernie's first bridge broke.

Explanation: Be Bernie and explain why your final bridge was stronger than the others and allowed you to get home.

Explanation: Be an engineer and write a presentation entitled 'Why Is a Suspension Bridge Stronger than a Simple Beam Bridge?' You could use pictures or models in your presentation.

History

Research Victorian inventions and inventors, including Brunel.

Design/Technology

Set a bridge challenge. Build a bridge that must span a particular width and hold a certain mass.

Design and make a bridge prototype. Test it and evaluate it. Improve the design, record the final design and do a final test.

Find out what gives a tall tower stability. See who can make the tallest tower from blocks or from dry spaghetti with marshmallow joiners.

Sources and notes

Chapter 1 Introduction

Chapter 2 Introducing the Storytelling School approach

See Chris Smith and Adam Guillain, *The Storytelling School Handbook for Teachers* (Hawthorn Press, Stroud, 2014) for more on the Storytelling School approach and the various sources upon which this method draws, in particular the work of Pie Corbett (storytelling and literacy), Ben Haggarty (storytelling performance) and Alida Gersie (storytelling and learning).

Chapter 3 Biology stories

Topics 3.1 Seeds and seasons
Story: Anna's Apple Tree
This story was created by the authors.

Topic 3.2 Senses
Story: Going to Catch a Piggy-wig
This story was created by the authors.

Topic 3.3 Humans, healthy diets, teeth and bones
Story: Uncle jack
Story: The Smiling Princes
These stories were created by the authors.

Topic 3.4 Habitats
Story: Mummy, Can I Have a Penguin?
This story was created by the authors.

Topic 3.5 Care for the environment
Story: The Drop of Honey
This story is in many Arabian Nights collections, and variations of it can be found in Burma, Thailand and Iran.
Other print versions:
Margaret Read MacDonald, *Peace Tales* (August House, Atlanta, 1992)
Maung Htin Haung, Helen G. Trager and Paw Oo Tiset, *A Kingdom Lost for a Drop of Honey and Other Burmese Folktales* (Parent's Magazine Press, 1968)

Story: The Bird and the Forest Fire
This is one of the Jataka stories about the life of the Buddha. You can find it in any number of collections of Jataka and Buddhist stories. It's sometimes called 'The Brave Little Parrot' or 'The Brave Little Quail'. There is a very similar South American folktale that features a hummingbird.
Other print versions:
Rafe Martin and Susan Gaber, *The Brave Little Parrot* (G.P. Putnam's Sons, New York, 1998)
Michael Nicolli Yahgulanaas and Wangari Maathai, *The Little Hummingbird* (Greystone Books, Vancouver, 2010)
Dharmachari Nagaria and Sharon Tancredi, *Buddha at Bedtime* (Duncan Baird, London, 2010)

Topic 3.6 Health and germs
Story: The Story of Edward Jenner
Retold by the authors based on information from: www.bbc.co.uk

Story: The Broad Street Pump
Retold by the authors based on information from: www.choleraandthethames.co.uk

Topic 3.7 Flowering plants
Story: Jack and the Giant's Peach
Innovated by the authors from the English wondertale 'Jack and the Beanstalk'. Our main source was Kevin Crossley-Holland, *Folk Tales of the British Isles* (Faber & Faber, London, 1986). We find this old version more satisfying than later versions because the back-story of Jack's father makes sense of the rest of the plot.

Story: Emily's Bees
This story was created by the authors

Chapter 4 Chemistry stories

Topic 4.1 Uses of materials
Story: The Fairy Godmother's Day Off
Innovated by the authors from 'Cinderella'. This ever popular story can be found in Wilhelm Grimm and Jacob Grimm, *Complete Fairy Tales* (Routledge & Kegan Paul, Oxford, 2002). In Grimm collections it is often titled 'Ashputtel'. Teachers may also wish to read for comparison the old English story 'Mossycoat' in Alan Garner, *Book of British Fairy Tales* (William Collins, London, 1984).

Topic 4.2 Changing materials
Story: Death of a Pancake
Innovated by the authors from the Scottish folktale 'The Wee Bannock' as told by Gerry Durkin on: www.educationscotland.gov.uk

Topic 4.3 Rocks and fossils
Story: The Fossil Woman
Created by the authors from historical sources found on: www.bbc.co.uk

Topic 4.4 Properties of materials
Story: The Horses of Troy
Innovated by the authors from Greek mythology.
Other print versions:
Russel Punter and Matteo Pincelli, *The Wooden Horse* (Usborne, London, 2011)
Cari Meister and Nick Harris, *Wooden Horse of Troy* (Raintree, Basingstoke, 2012)
Saviour Pirotta and Jan Lewis, *Odysseus and the Wooden Horse* (Orchard Books, London, 2006)

Topic 4.5 The water cycle
Story: The Children of the Water God
This story was created by the authors.

Topic 4.6 Mixing and separating materials
Story: Nimblefingers
This story was created by the authors.

Chapter 5 Physics stories

Topic 5.1 Forces – pull
Story: The Giant Turnip
This is a traditional Russian story, retold by the authors.
Other print versions:
Irene Yates, *The Enormous Turnip* (Ladybird Books, London, 2012)
Katie Daynes and Georgien Overwater, *The Enormous Turnip* (Usborne, London, 2006)
Aleksei Tolstoy and Niamh Sharkey, *The Gigantic Turnip* (with CD) (Barefoot Books, Bath, 1998)

Topic 5.2 Light and night and day
Story: The East and the West
This story was created by the authors.

Topic 5.3 Sound and hearing
Story: Little Rabbit Goes Home
This story was created by the authors.

Story: The Blind Man and the Hunter
This is a popular traditional tale from Zimbabwe, retold by the authors.
Other print versions:
Alexander McCall Smith, *The Girl Who Married a Lion and Other Tales from Africa* (Random House, New York, 1989)
Hugh Lupton and Niamh Sharkey, *Tales of Wisdom and Wonder* (Barefoot Books, Oxford, 1998)

Topic 5.4 Forces – push, pull and twist
Story: The Wheel that Jack Built
This story was created by the authors.

Topic 5.5 Forces – magnetism
Story: The Magic Stone
This story was created by the authors.

Topic 5.6 Light
Story: The Torch
This story was created by the authors.

Topic 5.7 Electricity – series circuits
Story: The Lighthouse Keeper's Son
This story was created by the authors.

Topic 5.8 Sound
Story: The Bat who Learned to Click
This story was created by the authors.

Topic 5.9 Space
Story: Apollo 13
This story was created by the authors based on historical information found on the websites such as: www.space.com; www.nasa.gov; www.airandspace.si.edu

Topic 5.10 Electricity – parallel circuits
Story: The Rocket
This story was created by the authors.

Topic 5.11 Forces and bridges
Story: Bernie's Bridge
This story was created by the authors.

Storytelling Schools Series

At Storytelling Schools we are passionate about the power of storytelling to transform education and learning. When storytelling skills are systematically taught throughout the school, students can use the approach to build confidence and fluency in spoken language, and to raise standards of reading and writing. Storytelling also provides an engaging, inclusive and enjoyable approach to subjects across the curriculum. At Storytelling Schools we offer information, resources and training to teachers who wish to adopt this approach in their school. Hawthorn Press will publish the Storytelling Schools Series over the next two years in partnership with Storytelling Schools.

Please check **www.hawthornpress.com** for book publication dates. To order books in advance, please email **orders@booksource.net** or telephone (0845) 3700063.

Published Spring 2014

The Storytelling School: Handbook for Teachers
Storytelling Schools Series, Volume I
Chris Smith and Adam Guillain,
Foreword by Pie Corbett

This handbook describes a revolutionary way of delivering primary education. In a storytelling school all children learn to be storytellers, retelling and improvising stories from memory as a way of learning both language and subject content across the curriculum. Children graduate with a repertoire of their own stories to tell. This approach has been shown to raise standards and fire imaginations in schools throughout the United Kingdom.

Chris Smith and Adam Guillain show you how to make this happen in practice. The handbook draws on more than ten years' experience researching and developing this way of teaching. Piloted, researched and up-dated, this second edition includes new chapters on non-fiction teaching and cross-curricular integration across the school. It is packed with practical activities, examples, theory, charts, diagrams and photocopiable pages.

ISBN: 978-1-907359-38-5: 210 x 297mm
Ringbound: Paperback

147 Traditional Stories for Primary School Children to Retell
Storytelling Schools Series, Volume II
Chris Smith

This amazing resource is your one-stop-shop for inspirational primary school storytelling (age 5–11). This unique collection of tried-and-tested stories has a comprehensive set of indexes classifying the stories by age suitability, related topic, values, genre, plot type and country of origin, for easy reference and story selection.

The volume was created to help storytelling schools to plan their own story curriculum. However, it will be of interest and value to any primary school teacher wishing to incorporate oral storytelling into their teaching. Together with the companion volume, *The Storytelling School Handbook for Teachers*, it provides an essential reference for any school wishing to adopt the Storytelling School approach.

'I have been collecting stories matched up to common primary topics for the past ten years. These stories have all been told by me and then used by teachers. We know the stories work well, engage and inspire children. The collection is a simple one-stop-shop for primary school storytelling, allowing quick access and topic links. It takes ages otherwise to find the right story and work out whether it is good for telling.' Chris Smith

ISBN: 978-1-907359-39-2: 210 x 297mm
Ringbound: Paperback

Available Summer 2015

Science Through Stories: Teaching Primary Science with Storytelling

Storytelling Schools Series, Volume III

Jules Pottle and Chris Smith

The authors have created a collection of stories for children to retell as a springboard for primary science teaching. Some of the stories have been created to contain the content of the science curriculum. Others are traditional stories that fit well with science topics and are great for oral retelling. The authors explain how to tell the stories, suggest ways of using them as a starting point for science teaching and also show how to link them to the teaching of literacy. Intended primarily as a resource for Storytelling Schools, this book will be of interest to all primary teachers who are looking for new ways to engage and inspire their classes about science.

ISBN 9789-1-907359-45-3: 210 x 297mm
Ringbound: Paperback

Available Spring 2016

History Through Stories: Teaching Primary History with Storytelling

Storytelling Schools Series, Volume IV

Chris Smith and Adam Guillain

This is a unique collection of history-related stories, which can be learned and retold orally as a way of inspiring engagement and interest in history. The stories include commonly taught topics for primary school together with suggestions on ways to link the story to the teaching of history. The approach has been adjusted to fit with the 2014 primary national curriculum for England and Wales, and focuses mainly on British history from the end of the last ice age to the present day.

ISBN: 978-1-907359-44-6: 210 x 297mm
Ringbound; Paperback

Future publications

A Storytelling Approach to Primary–Secondary Transition

Storytelling Schools Series, Volume V

Chris Smith and Nanette Stormont

A smooth transition between primary and secondary school is an important and often neglected part of the educational system. At a time of major change children work with uncertainty and fears of change as they move to their new schools. For some a difficult transition may lead to disengagement from secondary education.

In 2013 a group of storytelling primary schools in Oxford explored a way of using the Storytelling Schools approach to help children navigate this change. The basic idea was to use a single story, the *Odyssey*, to form a bridge between the two worlds: at the end of year 6 the students worked with the Cyclops story, using the Storytelling Schools methods and exploring issues of change, uncertainty and fear in the story and in relation to the move to a new school. Then at the beginning of year 7 students worked with the end of the story, *Return to Ithaca*, using a similar learning scheme. In this way there was a sense of continuity and consistency between the two schools.

The authors have produced a step-by-step guide to adopting this model. While it is designed with Storytelling Schools in mind, it is a model which can be adapted for all primary–secondary transitions.

Model Primary Storytelling School Scheme

Storytelling Schools Series, Volume VI

Nanette Stormont and Chris Smith

The Storytelling School approach to education uses oral storytelling as a springboard for learning language, raising standards in writing and teaching in almost any topic of the primary curriculum. This scheme provides a detailed example of a whole Storytelling School system, featuring:

- traditional and fiction stories for retelling;
- sample plans for teaching narrative writing;
- links to reading;
- non-fiction topics and texts;
- sample plans for teaching non-fiction using the storytelling approach;
- sample plans for cross curricular links to other subjects;
- tools for monitoring, evaluation and supervision;
- communications materials for staff, students and parents to explain and support adoption of the approach.

It is intended as a reference for schools wishing to become Storytelling Schools, as a blueprint to be imitated and innovated as needed.

The Storytelling School, Early Years Handbook
Storytelling Schools Series, Volume VII
Chris Smith and Adam Guillain

Storytelling and storymaking are an integral feature of early years education, providing a crucial spur for child development in those first few years. Exposure to a rich storytelling environment builds language and social development in a natural and enjoyable way.

In this handbook the authors provide a step-by-step guide to developing storytelling in these settings, with guidelines for working with children from birth to 5 years. Areas include:

- ways of telling stories to children;
- teaching the children to retell stories;
- group storymaking and retelling;
- the storytelling corner;
- storymaking in free play;
- storytelling and communication with babies and toddlers;
- using chants and ritual to support the rhythm of the day.

Stories, Chants and Rhymes for Early Years Children to Retell
Storytelling Schools Series, Volume VIII
Chris Smith and Adam Guillain

Every early years practitioner needs a repertoire of stories, songs, chants and rhymes so that they can create a rich story environment in their setting. In this volume the authors have collected a set of wonderful stories and rhymes that can be used to build the repertoire of every early years worker.

Index

A

Absorbency 123
Air 177
Air molecules 211
Air plant 91
Air resistance 233, 235
Aluminium 184
Amphibian 68
Antarctica 49–53, 55, 64
Apollo 13 3, 213-215
Apples 21–25, 88, 177
Appliance 202, 204
Archimedes screw 177
Archippos 116–124
Arrow 121, 147, 154, 163, 166
Astronauts 214, 217, 218, 220
Atmosphere 153, 210, 215, 217, 218, 223
Autumn 23, 24, 27, 58, 88

B

Bacterium 77
Ball 105, 112, 147, 153, 218, 219
Balloon car 177
Bandaging 45
Basalt 112
Bat (animal) 3, 207–210, 212
Battery viii, 4, 5, 190, 191, 202–204, 215, 223–226
Beam (light) 190, 193, 194
Bean 27, 52, 82, 83, 92, 154
Beanstalk viii, 28, 81, 83–85, 90, 92, 100, 238
Beehive 51, 88
Bees 1, 24, 49, 51, 53, 55, 67, 83, 87–90
Bellows 103, 172–175
Bicarbonate of soda 112, 130, 132, 141
Bingo 54
Bird 1, 5, 6, 54, 55, 63–68, 88, 138, 157, 158, 160, 166, 167, 169, 180
Blacksmith 103, 171–174, 177
Blind 3, 34, 163–167, 169, 195, 239
Bloodhound Project 235
Blowing 146, 147, 154
Bluebottle 59
Boats 147, 176
Boil (spot) 72, 73
Bones 1, 29, 37, 41, 46, 109, 212, 238
Bow 166, 169, 232
Brain 46
Brick 98, 99, 124, 176
Bridge ix, 3, 12, 15, 17, 229–234, 236, 239
Brighter 202, 204, 225, 226
Broad Street 1, 75–77, 80, 238
Brunel 233, 234, 236
Bulb (light) 5, 191, 195, 199, 202–204, 221, 224–227
Bulbs (plant) 91
Butter 2, 101, 102, 104–106, 131, 141, 142
Buzzers 202, 226

C

Cacti 91
Candle 105, 132
Car 176, 177, 222, 225, 227
Carbon dioxide 132, 217, 227
Cardboard viii, 91, 98, 100, 211
Carpenter 108, 171–175
Carriage 93–100
Cart 95, 96, 230–233
Cat 57, 59
Cells 202, 204, 226
Celsius 133
Chalk 100, 112, 154
Chocolate 33, 42, 43, 131, 142, 223
Choice chamber 68
Cholera 75, 76, 77, 78, 238
Choral 11, 154
Cinderella 93–100
Circuit 3–5, 14, 195, 197, 201–205, 221, 224–226, 233, 239
Circuit diagrams 202, 203, 226
Clay 99, 105, 106, 112, 115, 120–123, 147, 148
Click 3, 207–210, 239
Clouds 83, 125, 127, 128, 133
Coffee 38, 58, 141
Cold 6, 7, 23, 33, 52, 79, 109, 123, 125, 127, 131, 132, 150, 151, 207, 208, 215, 223, 232
Combe Mill 171
Compass 3, 154, 155, 179, 184, 185
Compression waves 211
Condensation 125, 130, 131, 140
Conductor 197, 202, 203, 226
Cough 207–210
Cow 6, 73, 82, 83, 92, 230
Cowpox 71, 72,
Cress seeds 154
Crops 91, 126, 132, 150, 151, 154
Crown 135, 137–140, 142
Crystals 132, 137, 141
Cup telephones 168
Cure 72, 73, 80
Curiosity – the Mars rover 219

D

Dark 6, 27, 117, 150–155, 192, 215
Dark box 193
Dark den 154
Data 68, 79, 80, 133. 184, 185, 226
Day 3, 148–154
Decay 28, 79
Deepening (in storytelling) 7–9, 16, 17, 19
Dentist 45, 47
Diet 1, 37–47
Digestive tract 46
Dimmer 202, 204, 225, 226
Disease 1, 6, 72, 63, 76, 80

Dissolve 46, 59, 109, 132, 140, 141
Doctor 35, 45, 47, 71–73, 75, 76, 78, 79
Dodo 70
Dog 57–61, 144, 145, 168
Dolores 94–100
Dough 79, 105, 147
Drag 235
Duck 49, 51, 53, 55, 157, 158, 160

E

Eagle god 63, 65, 66
Ears 34, 35, 47, 160, 164, 167–169, 207, 209, 210
Earth 107, 128, 149–153, 155, 169, 183, 184, 217–220, 222, 225, 233
Ear trumpets 168
East 3, 58, 60, 149–152, 155, 239
Echoes 207, 209
Echolocation 207, 210, 212
Ecosystem 1, 6, 57, 63, 68, 69
Edward Jenner 1, 71–73, 80, 238
Eggs 6, 46, 88, 102, 105
Electrical resistance 225, 226
Electricity 3–5, 14, 57, 197, 199, 202–205, 221, 225, 226, 239
Endangered species 57, 66, 69
Engineers 215, 236
Environment ix, 1, 53, 55, 57, 63, 66–70, 238, 242
Evaporate 66, 125, 127, 131, 132, 137, 141
Evaporation 13, 125, 130, 131, 135, 140
Evidence 75, 77–79, 184, 207
Evolution 3, 92, 207, 210
Exercise 79
Extinction 57, 63, 68–70, 112
Eye 34, 35, 47, 115, 117, 123, 153, 169, 192, 194, 208, 209

F

Fainter 167, 210
Fever 72, 73
Filter 135, 140–142
Fire 1, 27, 63–66, 99, 123, 174, 238
Fish 50–52, 54, 64, 68, 108, 160, 176, 184
Flexibility 93, 99, 115
Flies (insect) 68, 207–209
Flooding 133
Flour 102, 105, 131, 141, 172, 174, 211
Flower 1, 6, 24, 33, 51, 68, 81, 83–92, 238
Food chain 1, 53, 54, 57, 63, 67, 68, 169
Football 47, 68, 93, 96, 99, 147
Force 3, 143, 146, 147, 171–176, 179, 183, 184, 218, 229, 233–235, 239
Forest 1, 6, 63–66, 160, 164, 165, 167, 179–183, 238
Forge 103, 172, 175
Fossils 2, 107–113, 239
Fox 54, 101, 103, 104, 106
Friction 171, 175, 176, 235
Fruit 21, 26, 28, 34, 81, 83, 84, 85, 91, 92
Funnel 112, 141

G

Galileo 235
Gas 13, 130–132
Gears 175
Generate/reflect/select 16–17
Generator 197, 199, 201, 202, 204
Geologist 111
Germs 1, 71, 75, 79
Giant 27, 81–85, 90, 92, 143, 144, 238, 239
Global warming 57, 69, 133
Gloop 105, 131
Gold 135, 138, 140, 142
Golden stone 81, 83, 84
Granite 112
Gravity 3, 146, 176, 177, 215, 217–219, 233
Grind 172, 174, 175
Grip 176
Ground Control 213–215, 217, 220

H

Habitat 1, 3, 49, 53–55, 67, 68, 70, 141, 157, 169
Hardness 112, 122
Harvest 28, 154
Health 1, 26, 35, 37, 38, 41, 46, 47, 71, 74–80, 153, 155
Hear 3, 4, 5, 10, 14, 18, 30, 33, 34, 157, 163, 165–169, 211, 212
Hear/Map/Step/Speak 5, 7, 8, 14, 16, 17, 19, 26, 33, 45, 53, 66, 78, 90, 98, 104, 111, 122, 130, 140, 146, 153, 167, 175, 183, 192, 202, 210, 218, 2225, 233
Heart 6, 38, 46, 79
Heart rate 79
Heater 225–227
Honey 1, 51, 57–60, 66, 67, 69, 87–89, 238

I

Ice lollies 105, 131
Ichthyosaur 108, 111, 112, 113
Igneous 112
Impermeable 112
Infect 73, 109
Innovate 10, 12, 18, 28, 35, 47, 55, 69, 92, 100, 106, 124, 132, 142, 148, 155, 169, 177, 185, 195, 205, 212, 220, 227
Instruments 35, 167, 168, 212
Intestine 46
Invent viii, 8, 9, 10, 13, 16–18, 100, 115, 116, 122, 148, 185, 212
Invertebrate 68
Iron 101, 103, 172, 173, 179, 185
Irreversible 130, 132

Island 14, 137, 138, 197, 198, 199, 201, 202, 205, 227

J

Jam 101–104, 106
Jelly 105
Joints 46

K

Kites 147

L

Lambs 231–233
Lamp 4, 5, 153, 197–199, 201, 202, 227
Laser 194, 195
Lathe 172–175
Leaves 27, 28, 54, 90, 91, 124
Leg 34, 68, 79, 80, 102, 103, 121, 158, 234
Lens 112, 198, 199, 201, 203, 223
Levers 175
Levitating 184
Life cycle 1, 6, 21, 26–28, 81, 87, 90, 92
Light viii, 4–6, 26, 90, 149–155 187, 190–195, 203, 223–227
Lighthouse 3–5, 14, 195, 197–199, 201, 202, 204, 205
Light source 154, 187, 192, 193
Limestone 112
Liquids 104, 131, 140
Little Red Hen 154
Living 53, 54, 67, 68, 71, 75, 90, 109
Lodestone 179, 182–185
Log pile 54
Louder 157, 160, 211
Lunar module 215, 217
Lung 4, 38, 79

M

Magical viii, 8, 94, 184
Magic beans 28, 82, 92
Magnet viii, 146, 147, 176, 179, 183–185
Mammal 68
Marble 112
Marble run 176
Marshmallow 105, 132, 236
Mary Anning 113
Mass 99, 123, 176, 219, 234
Materials 93, 101, 115, 135
Medicine 71, 75, 78, 79
Melt 4, 101, 102, 104, 105, 117, 131, 138, 140, 142
Mendel 92
Metal 98, 99, 103, 115, 118, 122, 184, 211, 220
Metamorphic 112
Meteor 214
Microbes 71, 75, 78, 79
Milkmaid 71–74, 102, 103
Miller 72, 171–175
Millstone 172–175
Minerals 109
Minibeast 54, 68
Molecules 130, 211
Moon viii, 3, 8, 187, 191–193, 213–215, 217–220, 222, 223, 225
Morse code 195
Motorcar 100, 147
Mould 6, 105, 117, 121, 138
Muscles 6, 45, 59, 79

N

Nails 103, 104, 112, 232
Natural selection 210
Navigate 185, 241
Neil Armstrong 220
Newton 105, 176, 177
Newton metre 176
Night 4, 149, 152–154, 195, 218, 219
Nose 34, 227
Nutrient 46, 90

O

Obsidian 113
Ocean 126–128, 130, 133, 185
Odysseus 115–121, 124
Oesophagus 46
Oil 79, 141, 174, 175, 177, 198, 201, 202
Opposing forces 233
Optic fibre 194
Orbit 149, 153, 213, 217–219, 223, 225
Orchard 88, 89
Orrery set 219
Ovary 91
Owl 157, 160
Oxygen 79, 213–215, 218

P

Paddles 172–174
Palaeontologist 107, 110
Pancake 101–106
Paper 46, 80, 98, 99, 106, 112, 113, 115, 123, 124, 141, 154, 168, 194, 211, 232, 234
Paper clip 184, 203
Papyrus 117, 123, 124
Parallel circuits 197, 201, 204, 221, 224–226
Particles 112, 131, 141
Particle theory 131
Pasta 105, 141
Peaches 81, 83, 84, 90, 92
Penguin 49–55, 64, 176
Penicillin 80
Periscope 193, 195
Photosynthesis 91
Pillar 232, 234
Pinhole camera 194
Pitch 168, 210–212
Pitfall trap 169
Planet 150, 218–220, 222
Plaster of Paris 141
Plastic 38, 54, 68, 98, 99, 111, 154, 168, 169, 185
Playdough 105
Pneumatic 203
Polar bear 55, 135, 136, 139
Poles 183, 184
Pollen 83, 88, 89, 91

Pollination 67, 81, 85, 87, 90, 91
Pond 51, 54, 68
Pop rockets 227
Porous 112
Predator 67, 169
Prehistoric 107, 108, 110, 113
Preserve 79, 109
Prey 34, 54, 67, 169, 212
Prince 41, 42, 44, 45, 47, 94, 96, 97, 116
Princess 41, 42, 43, 44, 45, 47
Professor 221–227
Prototype 236
Pull 68, 69, 143–148, 171, 175, 176, 217, 218, 225
Pump 75–78, 80, 172–175
Pumpkin 93, 96, 97, 100
Push viii, 143, 144, 146–148, 154, 169, 171–177, 183, 235

R

Rabbit 49, 50, 53–55, 157, 158, 160, 167, 195
Rain 99, 125, 128, 133
Rain gauge 133
Rationing 47
Recipe 28, 35, 105, 106, 148
Recycled paper 124
Reeds 211
Reflected (light) 192–195
Reflected (sound) 207, 210
Reflector 187, 191–193
Repelling 147, 183, 184
Reproduction 1, 90
Reptile 68, 107, 108, 113
Resistance 204, 225, 226
Reversible 130–132, 140
Ring magnet 179, 184
Ripe 81, 84
River 61, 65, 76, 127, 128, 133, 138, 158, 160, 165, 171, 172, 175, 230–233
Rock 4, 98, 99, 107, 109–113, 122, 123, 199, 201, 231
Rocket 214, 217, 218, 221–223, 225, 227
Roots 27, 91, 144
Rope 202, 232
Rotating 153, 219
Rotation 149, 218, 219
Rules 35

S

Sails 147
Salt 79, 132, 135, 137, 139, 141, 180
Sand 105, 109, 111, 112, 117, 123, 141
Sandpaper 91
Sandstone 112
Sandwich 34, 35
Scent 33, 59, 91, 165
Seashells 107
Seasons 21, 24, 26–28, 154, 219
Sedimentary 111, 112
Seed 21–24, 26–28, 81, 84, 88, 90–92, 96, 144, 154
Senses 6, 29, 33–35, 104, 163, 167
Series circuits 197, 203, 204, 225, 226
Seven basic plots 8
Sewage 76, 77
Shadow 154, 191–193, 219
Shadow puppet 193
Shaduf 177
Shale 112
Sharing 155
Sheep 229–233
Shepherd 57–61, 66
Shoots 65, 144
Sieve 135, 138, 141, 142
Sight 34, 163, 167, 208, 209
Signal 187, 190, 191, 192, 194, 195
Sign language 169
Skeleton 45–47, 107–109, 111
Skull 46, 59, 109
Slate 112
Smallpox 71–74, 78, 80
Smell 30, 33, 34, 41, 42, 83, 84, 89, 102, 163, 167, 180, 181
Smile 41–44
Snail 157, 160
Snake 55, 165
Sneeze 208
Soil 26, 27, 90, 109, 111, 112, 122, 123, 132, 138, 141
Solar system 149, 155, 218, 227
Soldiers 60, 117, 118, 120, 188
Solids 104, 131, 140, 141
Solubility 122
Sonar 212
Sound 16, 33–35, 66, 157, 160, 161, 163, 167–169, 181, 207, 209–212
Soup 146, 148
Space 152, 155, 168, 213–215, 217–223, 225, 227
Spacecraft 213–215, 218
Space food 220
Spaceship 217, 218, 220
Space travel 213, 220, 227
Spinners 235
Spinning 136, 149, 152, 173, 174, 219
Spray 89, 90, 131, 133
Spring 176
Spring (season) 23, 24, 27, 28, 58, 65, 88
Squeaker 211
Stability 236
Stars 185, 222, 223, 227
Steel 112, 184, 185
Stem 90, 91
Stigma 91
Stomach 46
Stone 14, 81, 83, 84, 121–124, 138, 179–183, 232
Storytelling and writing 16
Streamlining 175, 176
Strength 17, 93, 99, 100, 115, 123, 161, 184, 185, 232–234
Sugar 105, 123, 131, 141
Summer 22–24, 27, 230
Sundials 219
Sun God 149–153, 155
Support 45, 46, 91, 213, 234
Suspension bridge 232–234, 236

Switches 4, 5, 197, 201–203, 214, 226
Symmetry 55

T

Taste 31, 33, 34, 47, 58, 59, 137, 146, 167, 209, 220
Teaching sequence 7, 9, 17
Tears 63, 65, 96
Teeth 35, 37, 41–43, 45–47, 59, 109
Telephone 168, 194, 212
Telescope 222, 223
Temperature 73, 123, 133, 153, 154, 219
Terminal 203
Theory 71, 75, 78–80, 207, 210, 235
Thermal conductance 115
Thought corridor 66, 78, 130, 167, 183, 192, 202
Time-lapse 27, 91, 154
Tools 122, 230, 242
Torch 154, 187, 188, 190–193, 195
Transparency 115, 122, 123, 193, 194
Trap 166, 169
Tree 21, 22, 24, 26–28, 64–66, 69, 81, 83–85, 90, 91, 108, 154, 177
Trick 12, 17, 120, 163, 182
Troy 115, 116, 120, 122, 124
Tug of war 147, 233
Tuning fork 211
Tunnel 187, 190, 191, 192, 195
Turnip 28, 143–146, 148
Twist 98, 104, 146, 150–152, 171, 175, 176

V

Vaccine 71, 73
Vegetables 34, 148, 154
Vibrations 210–212, 214
Vinegar 112, 130, 132, 141
Volcano 112, 113
Volume 210, 211

W

Washing 46, 80
Water ix, 13, 26, 27, 34, 46, 66, 75–79, 90, 91, 98, 105, 112, 115, 123, 125–128, 130–139, 141, 153, 154, 168–177, 203, 211, 211
Water cycle 13, 125, 130–133
Water god 125, 126, 128, 130, 132, 133
Waterproof 93, 95, 96, 99, 115, 118, 121, 123
Water resistance 176
Water vapour 125, 127, 130
Water wheel 171, 173–175, 177
Wax crayons 105
Weather 26, 27, 95, 96, 130, 133
Weight 91, 99, 105, 106, 115,
169, 177, 232–236
Wheat 89, 172
Wind dispersal 91
Wings 34, 64, 65, 158, 208, 210, 235
Winter 23, 24, 27, 28, 131, 154
Wire viii, 4, 5, 14, 197, 199, 201–205, 222, 224–227, 233
Wolf 6, 7, 12, 29, 30, 33, 35, 169
Wood 54, 68, 93, 95, 98–100, 115, 116, 118–120, 122, 124, 138, 172, 176, 211, 231, 232
Wooden horse 116, 119, 124
Woodlice 54, 68
Wormery 112

Y

Yeast 79

Other Books from Hawthorn Press

The Islamic Year

Surahs, stories and celebrations

Noorah Al-Gailani, Chris Smith

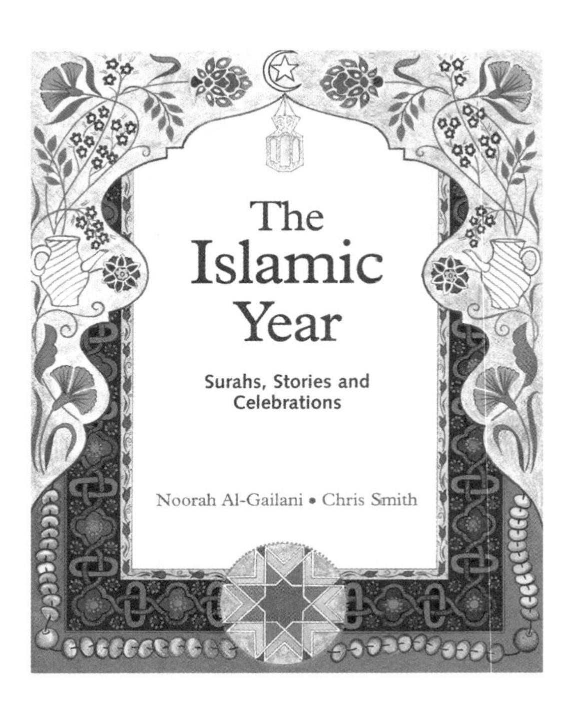

Celebrate the Islamic Year in your family or at school with this inspiring treasury of stories, surahs, songs, games, recipes, craft and art activities. Folk tales illustrate the core values of Islamic culture with gentle humour and wisdom. *The Islamic Year* is beautifully illustrated, with a calendar, Arabic calligraphy of the Names of God, traditional patterns, maps and pictures drawn from many part of the Muslim world.

'Educators seeking to interpret Islam to non-Muslim audiences will ind this an excellent resource.'

Baroness Uddin, House of Lords, Westminster

ISBN: 978-1-903458-14-3; 250 x 200mm; paperback

The Storyteller's Way

A Sourcebook for Inspired Storytelling

Sue Hollingsworth, Ashley Ramsden

Everyone can tell a story, but to tell it well you need a certain set of skills. Whether you're starting out or want to develop your storytelling expertise, this book is an essential guide.

Use it to tell stories for entertainment, teaching, coaching, healing or making meaning. It contains a wealth of stories, exercises, questions, tips and insights to guide your storytelling path, offering time-tested and trusted ways to improve your skills, overcome blocks and become a confident and inspirational storyteller.

ISBN: 978-1-907359-19-4; 228 x 186mm; paperback

Jumping Mouse

Brian Patten

Illustrated by Mary Moore

This Native American folk tale tells how a little mouse had the courage to leave his family's warm nest in the roots of a giant tree, and go in search of the Sacred Mountains. He has many adventures on the way and finds his true name of Jumping Mouse. This hero quest has enchanted generations of children since first publication in 1972. Brian Patten has updated his original version of the story for this second edition, which also reproduces the original illustrations by Mary Moore, the daughter of the sculptor Henry Moore.

'A small masterpiece.' Charles Causley

ISBN: 978-1-903458-99-0; 210 x 210mm; hardback

The Natural Storyteller

Wildlife Tales for Telling

Georgiana Keable

Georgiana Keable introduces us to a staggering wealth of world stories all about nature and our role as humans in it. Culturally diverse and all told with great energy and panache, the stories will engage young readers and encourage them to become natural storytellers. The book includes several storymaps, and each section has a practical activity that can be undertaken individually or as a group.

The author's message is clear: the resources needed for Natural Storytelling are abundantly around us – nature and our imagination.

ISBN: 978-1-907359-80-4; 228 x 186mm; paperback

All the Dear Little Animals

Ulf Nilsson

Illustrated by Eva Eriksson

Three children decide someone must bury all the world's poor dead animals:

'The whole world is full of dead things,' said Esther. 'In every bush there is a bird, a butterfly, a mouse. Someone must be kind and look after them. Someone must make a sacrifice and see that all these things are buried.' 'Who must?' I asked. 'We must,' she said.

This is a picture book for children aged five and above. It covers a difficult subject in a wonderfully unsentimental way, describing exactly how children can resolve big issues – through play.

'This captivating book takes us on a safe, funny and deeply meaningful adventure.' Julie Stokes OBE, Founder and Clinical Director of Winston's Wish ISBN: 978-1-903458-94-5; 210 x 218mm; hardback

Storytelling with Children

Nancy Mellon

Telling stories awakens wonder and creates special occasions with children, whether it is bedtime, around the fire or on rainy days. Encouraging you to spin golden tales, Nancy Mellon shows how you can become a confident storyteller and enrich your family with the power of story. Find the tale you want from Nancy's rich story-cupboard.

Nancy Mellon runs a School for Therapeutic Storytelling and lives in New Hampshire.

ISBN: 978-1-903458-08-2; 216 × 138mm; paperback

Writing to Reading the Steiner Waldorf Way

Foundations of Creative Literacy in Classes 1 and 2

Abi Allanson and Nicky Teesma

The book showcases the holistic, creative aspects of the Steiner Waldorf literacy approach. The teaching of writing before reading is prioritised so as to engage children's creativity in learning. Developing the child's own voice through writing and storytelling, to lead over into reading, is highly effective for motivation and success.

ISBN: 978-1-907359-88-0; 246 x 189mm; hardback

An A-Z Collection of Behaviour Tales

From Angry Ant to Zestless Zebra

Susan Perrow

Time tested, imaginative stories to help children with life challenges and behaviour issues.

Following the alphabet from A to Z, each story in this collection begins with an undesirable or out-of-balance situation and, through the use of metaphor and an imaginative story journey, leads to a more desirable resolution.

The stories can be told directly, or adapted. They can be turned into home-made picture books and puppet shows, or used as springboards for the creation of new tales for particular behaviour challenges and situations.

ISBN: 978-1-907359-86-6; 234 x 156mm; paperback

Ordering Books

If you have difficulties ordering Hawthorn Press books from a bookshop, you can order online at **www.hawthornpress.com** or order direct from

BOOKSOURCE

50 Cambuslang Road, Glasgow G32 8NB Tel: (0845) 370 0067 orders@booksource.net

Hawthorn Press

www.hawthornpress.com

LUCET BRAIDING

VARIATIONS ON A RENAISSANCE CORD

by ELAINE FULLER

Illustrations by SORREL SMITH

LACIS
PUBLICATIONS
Berkeley, CA USA

DEDICATION

Frank C. Gardiner in memoriam

ACKNOWLEDGMENTS

I would like to thank Viola Buckner, Kristine Comito, Anne Felden, Cynthia Lynn Hussey, Carole Owens, Persephone Keckley, and Julie St. Germaine for testing the cord directions. I would also like to thank Jules and Kaethe Kliot for their support and encouragement, and Perrin Kliot for computer help and photography. Finally, I would like to thank James and Julie for being patient.

LACIS
PUBLICATIONS
3163 Adeline Street
Berkeley, CA 94703

ISBN 978-1891656-06-4

CONTENTS

I. Introduction 3
II. Terms & Abbreviations 4
II. Single thread cords 5
Basic Single-Thread Cord 5
Single-Thread Cord: No-Turn Method 7
Single-Thread Cords with Twisted Stitches 8
III. Multiple thread cords 11
Cord with One Gimp 11
Cord with Two Gimps 12
Cords with Thread Exchange 13
Three Color Cord With Thread Exchange 14
Cords with Two & Three Working Threads 15
IV. Beaded Cords 23
V. Left-Handed Lucet 26
Left-Handed Single-Thread Cord 26
Left-Handed No-Turn Cord 27
Right and Left Twist Stitches for Left-Handers 29
VI. Troubleshooting 30
VII. Applications 31

INTRODUCTION

The lucet is a simple tool used for making cords and drawstrings. The earliest lucet yet recovered dates to the Viking era, and this little implement was in common use in Europe from at least the 16th century until the advent of cheap, machine-made cords in the 1830's. The lucet experienced a brief revival in the mid-19th c., and again in the Edwardian era. Lucets were made of a variety of materials: bone, wood, and horn for the humbler workbasket, with fancier tools shaped from ivory, tortoise-shell or mother-of-pearl. Some of the finest surviving examples have metal inlays. Steel inlays were particularly favored in the final years of the 18th c., reflecting the late Georgian fashion for steel-thread embroidery on costume.

Nearly any strong, even thread or yarn can be used for lucet cord. More fragile threads, such as softly twisted wool, chenille, or silk floss, can be used as a gimp thread, as can ribbons and metallics. Number 3 pearl cotton, bedspread weight crochet cotton, or even heavy string are good for the first practice sessions. If no specific thread is mentioned in the instructions, use one of these three. Later cords can be worked in finer threads. Early 19th c. workers appear to have favored silk for their cords.

The basic lucet cord uses a single thread to make a square braid. As you go through the techniques in this book, you will find unlimited possibilities as the basic cord evolves into multicolor braids, beaded trims and picot edgings for the finest costume.

The cords in this book are arranged in increasing order of difficulty. *It is important to work through the patterns in the order presented, as information given early on may not be repeated later.*

TOOLS, TERMS & ABBREVIATIONS

THE LUCET

The design of the lucet is simple: two fixed rods spaced about 2 inches apart. The two configurations are the "lyre" shape, without a handle, and the "Y" shape with an integral handle. The "lyre" shape was favored for lucets made of exotic or expensive materials. The "Y" shape, sometimes called a chain fork or hay fork, facilitates rotation of the lucet. Most lucets have a hole near the base of the uprights to guide the finished braid. The uprights will be referred to as "horns" in this text. The tops of the horns turn away from you and slightly up to keep the working loops in position.

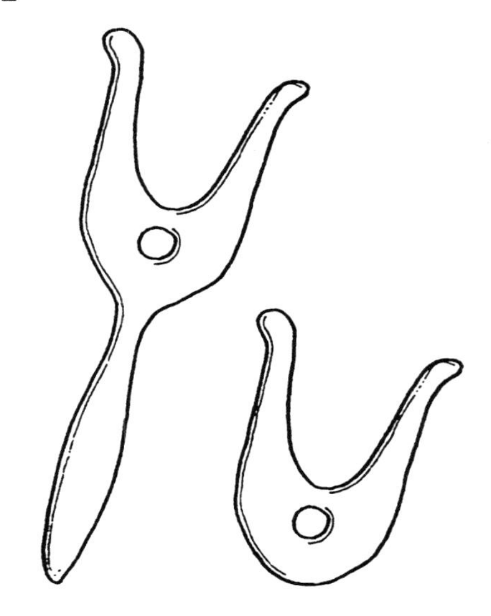

TERMS

Before you go any farther, please make sure you understand the following terms:

Base cord: The basic braid, formed by making stitches on the horns of the lucet.

Gimp: One or more supplemental threads weaving in and out of the base cord.

Knot: A completed stitch.

Kumi Himo Bobbin: A wooden spool used to store and tension thread for braiding.

Front: The side of the lucet facing you at any given time.

Stitch: The configuration of thread made by wrapping the thread around a horn of the lucet and lifting the old loop up, off, and over the horn.

Thread: Whatever material you are braiding, be it thread, yarn, string, ribbon, etc.

Turn: To turn the lucet over, 180 degrees clockwise if looking from above. If you are looking at the front of the lucet, it will appear to flip from right to left, as a page turns in a book.

Working Thread: The thread running between the lucet and the thread supply.

ABBREVIATIONS

B TO F	Back to front
DK	Dark
F TO B	Front to back
IFO	In front of
LH	Left-hand
LT	Light
MED	Medium
RH	Right-hand

SINGLE THREAD CORDS

BASIC SINGLE-THREAD CORD

The basic process consists of grasping loops and lifting them over the horn of the lucet. The left hand holds the lucet with the horns pointing away from you. The right hand controls and manipulates the thread.

1. Thread a six-inch tail of crochet cotton from F to B through the hole of the lucet. With your left hand, hold this tail securely against the back of the lucet. Wrap the thread from the ball twice around the little finger of your right hand, then pass the thread across the palm of your hand and hold it lightly between the right thumb and index finger.

2. Wrap the thread around the horns in a figure 8. Do this by taking the thread counterclockwise around to the back of the RH horn, bringing the thread forward between the horns, then wrapping the thread clockwise around the LH horn. Now bring the thread IFO the RH horn and *above the first wrap* (**Fig. 1**).

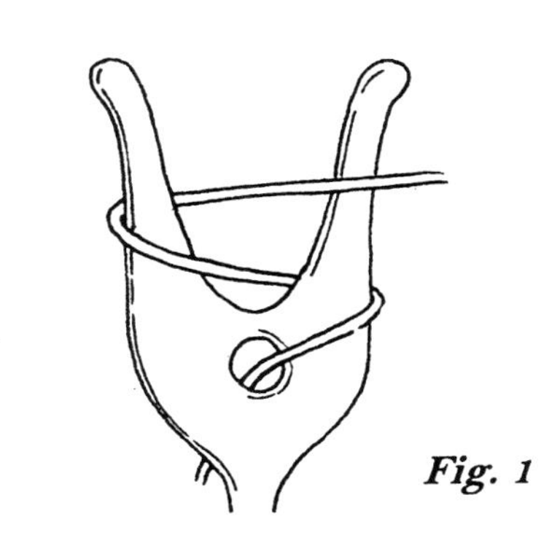

Fig. 1

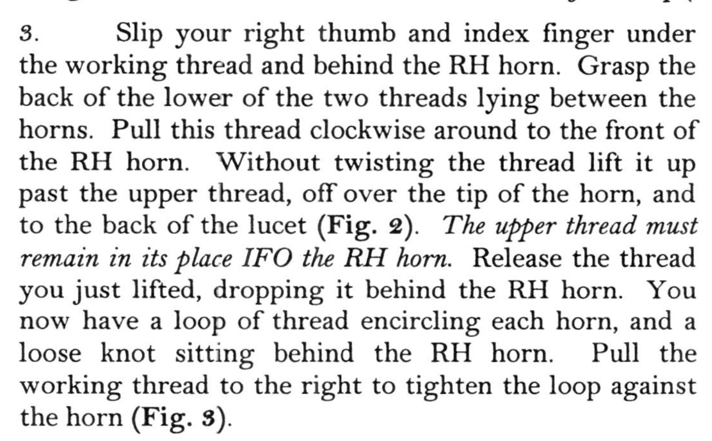

3. Slip your right thumb and index finger under the working thread and behind the RH horn. Grasp the back of the lower of the two threads lying between the horns. Pull this thread clockwise around to the front of the RH horn. Without twisting the thread lift it up past the upper thread, off over the tip of the horn, and to the back of the lucet (**Fig. 2**). *The upper thread must remain in its place IFO the RH horn.* Release the thread you just lifted, dropping it behind the RH horn. You now have a loop of thread encircling each horn, and a loose knot sitting behind the RH horn. Pull the working thread to the right to tighten the loop against the horn (**Fig. 3**).

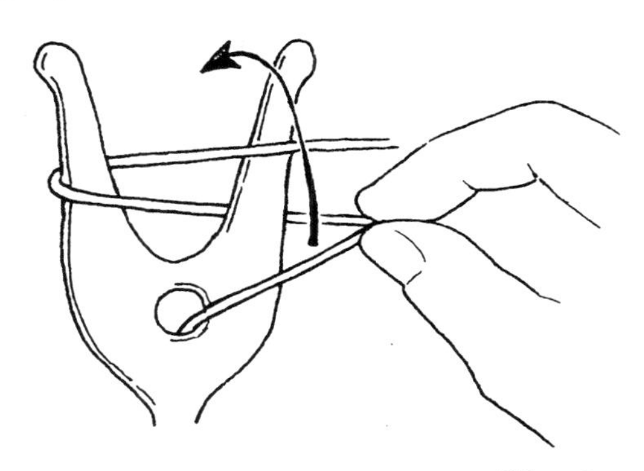

Fig. 2

4. Using your left hand, turn the lucet over from right to left (the lucet turns like a page in a book). The work should now appear as in **Fig. 4**, *with the working thread lying above the loop on the RH horn.*

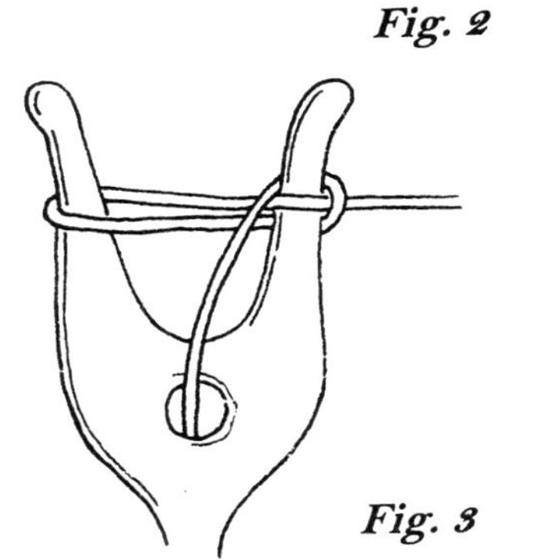

Fig. 3

5. Slip your right thumb and index finger under the working thread and behind the RH horn. Grasp the back of the loop on the RH horn. Pull this part of the loop clockwise around to the front of the RH horn. Without twisting the loop lift it up past the upper thread, off over the tip of horn, and to the back of the lucet (This is the same motion shown in **Fig. 2**). *The upper thread must remain in its place IFO the RH horn.* Release the loop you just lifted over the horn, dropping the loop to the back of the horn. As in step 3, you now have a loop of thread encircling each horn, and a loose knot sitting behind the RH horn. With your right hand, pull the working thread to the right to tighten the loop against the horn. Carry the thread across the back of the lucet, stopping just to the left of the LH horn. The work should appear as in **Fig. 5**.

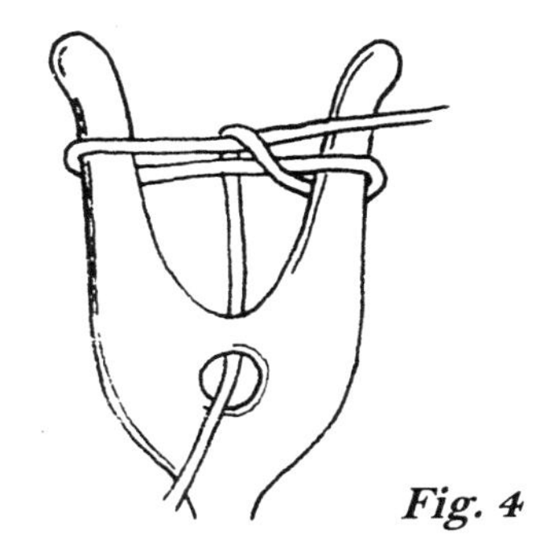

Fig. 4

REPEAT STEPS 4 AND 5.

If you have a square cord that looks like knit stitches or chain stitches on all four sides, congratulations! You've just taught yourself to make cord with the lucet.

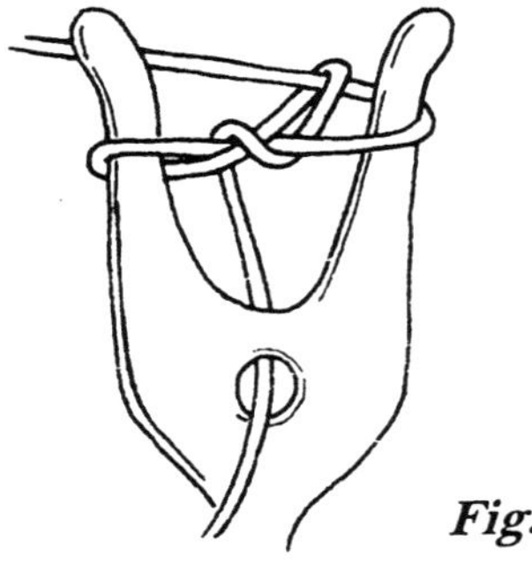

Fig. 5

As you become proficient, the individual hand motions involved in turning the lucet and positioning the working thread will flow into each other, as will the motions of grasping the loop and lifting it up, off, and over the tip of the horn.

A NOTE ON TENSION: *Beginners frequently experience difficulty in keeping the finished cord even. The key to a neat, smooth cord is to pull on the loop before you lift it off and over the horn. Pulling on the part of the loop that lies* **behind** *the horn loosens the loop so that it can be easily lifted. Pulling on the part of the loop that lies* **in front** *of the horn tightens the old knot and snugs it down into the cord. When you pick up the loop in step 5, pull first on the back of the loop to loosen it. Then pull on the front of the loop to tighten the previous knot neatly.*

*Holding the cord up near the loops in a pincer grip between the thumb and index finger of your left hand also helps keep the tension even. Some workers like to wind the working thread onto a kumi himo bobbin. Wind the bobbin and secure the thread as shown (**FIG. 6A and B**). Do not wrap the thread around the your fingers. Instead, turn the lucet and let the bobbin hang off the outer edge of the RH horn. When you need more thread, pull it to the left off the bobbin without undoing the hitch. This is a very fast way of producing a regular, even cord.*

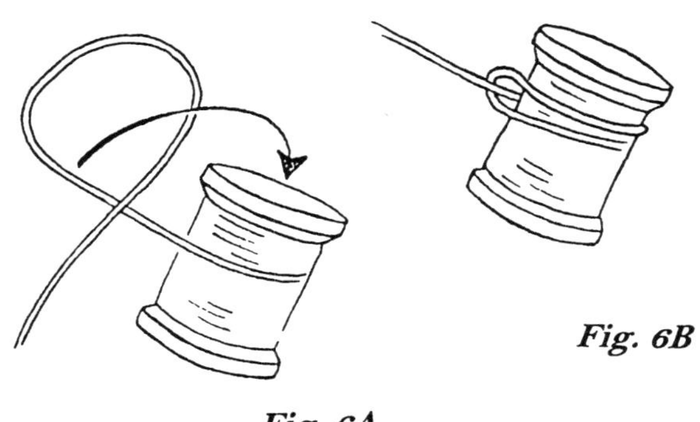

Fig. 6A

Fig. 6B

ENDING OFF THE CORD

1. Turn the lucet as though you were going to start a new stitch, but instead of lifting the loop, cut the thread, leaving about a six-inch tail.

2. Thread the tail down through the loop on the RH horn. Slip the loop off the horn, and pull on the thread until the loop closes.

3. Now thread the tail down through the loop on the LH horn. Slip the loop off the horn, and pull on the thread until the loop closes.

SINGLE THREAD CORD: NO-TURN METHOD

This method is particularly useful for advanced cords that require several strands and beads. Here, however, we will practice with a single thread. The position of the hands and the lucet is the same as for the basic cord.

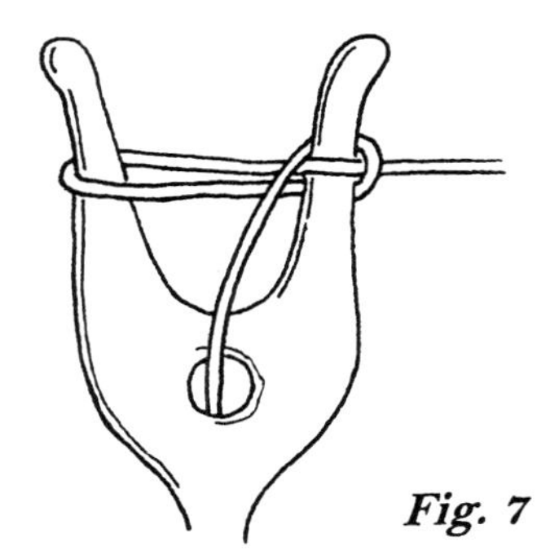

Fig. 7

1. Thread the crochet cotton and tension the ball thread as in step 1 of the basic cord. Wrap the lucet as in step 2 of the basic cord.

2. As in step 3 of the basic cord, grasp the lower of the two threads lying between the horns, pull it around to the front of the RH horn and lift it up, off, and over the tip of the horn, dropping it behind the RH horn. *The upper thread must remain in its place IFO the RH horn.* You now have a loop of thread encircling each horn and a loose knot sitting behind the RH horn (**Fig. 7**). Pull the working thread to the right to tighten the loop against the horn. Continuing to keep tension on the thread, move your hand to the left behind the lucet, stopping just to the left of the LH horn.

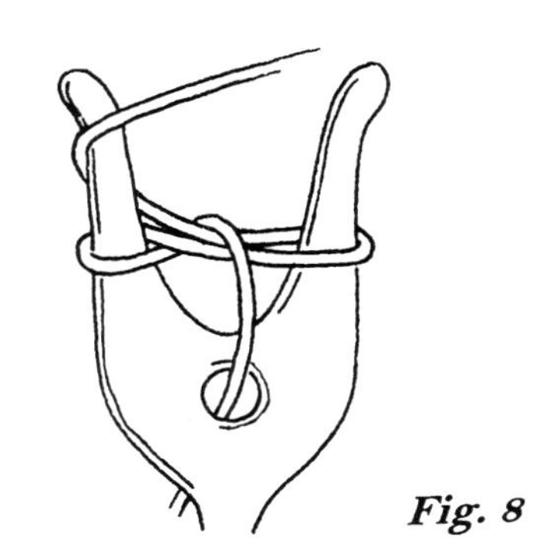

Fig. 8

3. Bring the thread counterclockwise around to the front of the LH horn. The new wrap lies above the loop on the LH horn (**Fig. 8**).

4. Grasp the part of the LH loop that lies IFO the LH horn and near the center of the lucet. Move your hand to the left, stopping when your thumb and index finger are even with the left edge of the LH horn. Lift the loop up past the working thread and off over the tip of the LH horn (**Fig. 9**). *The working thread must remain in its place on the LH horn.* Release the loop in the space between the horns. Pull the working thread to the right until the knot is centered between the horns. You may find it helpful to pull simultaneously on the tail with the middle, ring and index fingers of your left hand (your left thumb and index finger hold the lucet). *Leave the knot somewhat loose. You will tighten it later, in Step 6*

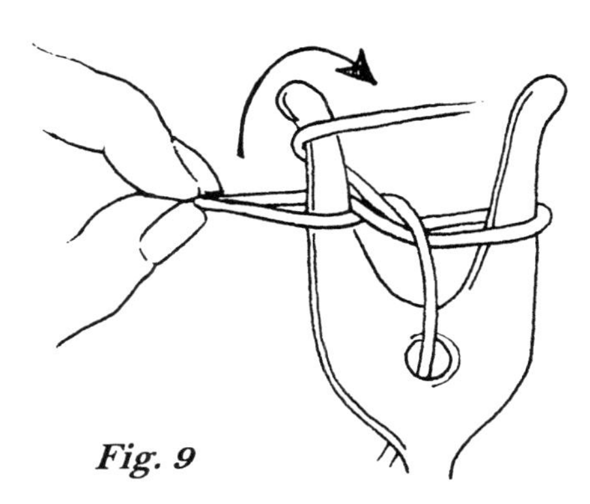

Fig. 9

5. Lay the working thread across and IFO the RH horn, with the thread above the loop.

6. Grasp the back of the loop on the RH horn and pull it around to the front of the RH horn. When your thumb and index finger have come all the way around to the front of the lucet, tighten the knot from step 4 by pulling on the LH half of the loop. Now lift the loop up, off, and over the tip of horn. Drop the loop behind the RH horn. *The working thread must remain in its place IFO the RH horn.* You now have a loop of thread encircling each horn and a loose knot sitting behind the RH horn. Pull the working thread to the right to tighten the loop against the horn. Carry the working thread to the left behind the lucet, stopping just to the left of the LH horn.

N.B.: In future cords, steps 5 and 6 will be summarized as "Make a stitch on the RH horn".

7. Bring the thread counterclockwise around to the front of the LH horn. The new wrap lies above the loop on the LH horn. **See Fig. 8 again.**

8. Grasp the part of the LH loop that lies IFO the LH horn and near the centered knots. Keeping hold of the loop, move your hand to the left, stopping when your thumb and index finger are even with the left edge of the LH horn. If necessary, pull on the back half of the loop to tighten the knot made in step 6. Lift the loop up past the working thread and off over the tip of the LH horn. This is the same motion shown in **Fig 9.** Release the loop and pull the working thread to the right until the knot is centered between the horns. For the first few stitches, you may find it helpful to simultaneously pull on the tail with the middle, ring and index fingers of your left hand (your left thumb and index finger are holding the lucet).

N.B.: In future cords, steps 7 and 8 will be summarized as "Make a stitch on the LH horn".

REPEAT STEPS 5 THROUGH 8.

When you have enough knots to hang on to, stop pulling on the tail. Instead, hold the cord up near the loops between the thumb and index finger of your left hand. The lucet will now rest between the palm and remaining fingers of your left hand. This helps keep the tension even.

Here's a little mnemonic to help you wrap the horns correctly: "Right in front" for wrapping the right horn F to B, and "left behind" for wrapping the left horn B to F.

ENDING A NO-TURN CORD

1. Stop after completing a stitch on the LH horn (Step 8). Cut the thread, leaving about a six-inch tail.

2. Thread the tail down through the loop on the RH horn. Slip the loop off the horn. Pull on the thread until the loop closes.

3. Now thread the tail down through the loop on the LH horn. Slip the loop off the horn. Pull on the thread until the loop closes.

SINGLE-THREAD CORDS WITH TWISTED STITCHES

Changing the way you place the working thread on the horns alters the appearance of the finished cord. Here, the working thread is placed in a half-hitch on the horn. This forms a twisted stitch which flattens the cord and makes it look braided. The flattened cord is especially suitable for trim. Rya yarn makes a handsome wool braid. Size 3, 5, or 8 pearl cotton are all nice. If you go much finer than size 8 pearl cotton, the "braided" interlacement of the loops becomes too small to see.

TURNED SINGLE THREAD CORD WITH TWISTED STITCHES

1. Thread a tail of crochet cotton from F to B through the hole of the lucet. The working thread lies in front of the lucet; pass it between the horns to the back. The working thread, now lying behind the RH horn, wraps twice around the little finger of your right hand, then passes across the palm of your hand. Fold your little, ring, and middle fingers down over the thread, and fully extend your right thumb and index finger.

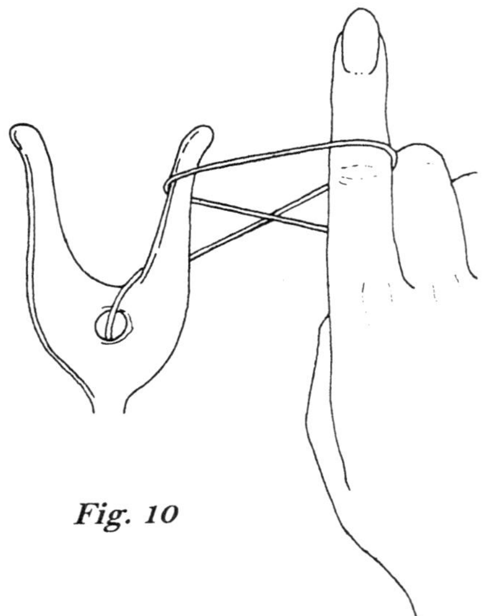

Fig. 10

2. With your finger going behind the working thread, turn your hand so that your index finger points down and your palm faces you. Bring your right index finger towards you to catch the working thread, and, tracing a semi-circle in the air, point first to yourself, then straight up, making a loop on your index finger. If the tip of your finger were moving around a clock face, it would go from 6 o'clock to 12 o'clock.

3. Keeping your index finger pointed up, slide the RH horn of the lucet from below along your finger into the loop (**Fig. 10**). Slip your finger out of the loop. This transfers the loop from your finger to the RH horn. Pull the working thread to the right to tighten the loop.

N.B.: In the future, steps 2 and 3 will be summarized as "Make a Right Twist Loop on the RH horn".

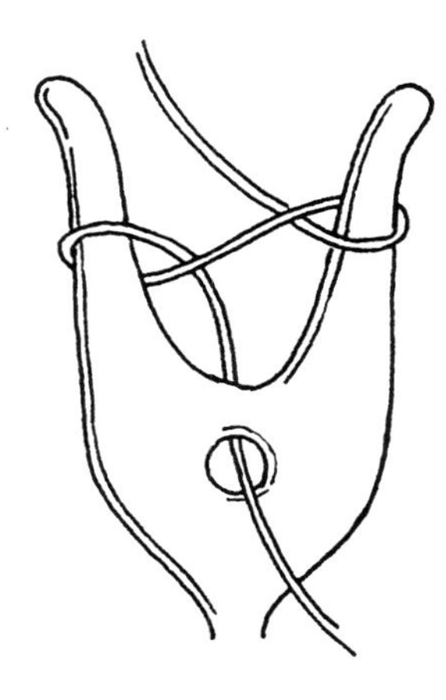

Fig. 11

4. Turn the lucet over from right to left. You have a half-hitch sitting on what is now the LH horn. Pass the working thread to the back of the lucet.

5. Make a Right Twist Loop on what is now the RH horn. You now have a half-hitch sitting on each horn (**Fig. 11**).

6. Turn the lucet over from right to left. Pass the working thread to the back of the lucet.

7. Make a new Right Twist Loop on what is now the RH horn and lift the old loop up, off, and over. Center the knot, *but be careful not to tighten it all the way. You will tighten this knot when you lift the next old loop up, off, and over.*

REPEAT STEPS 6 AND 7.

Regulating the tension of a twisted-stitch cord can be difficult at first. You must leave the knot in step 7 looser than in a regular cord so as to gain enough slack in the loop to lift it. A smooth thread like pearl cotton slides into place more easily than a hairy one and should be used for your first twisted-stitch cord. If the loop to be lifted gets too tight and simply refuses to budge, try carefully pulling on the back of the other loop and then on the front of the loop to be lifted. If, on the other hand, the knot is so loose as to make the cord irregular, try pulling the working thread to the right after you have turned the lucet but before lifting off the loop.

ENDING OFF THE CORD

When you are finished with this cord, end off as usual for the basic, turned cord.

SINGLE THREAD CORD WITH TWISTED STITCHES: NO-TURN METHOD

1. Thread a tail of crochet cotton and tension the ball thread as in step 1 of the Turned Cord with Twist Stitches. Make a Right Twist Loop on the RH horn.

2. Point your left index finger up. Moving you right hand from right to left, carry the working thread across and IFO your left index finger. Wrap the working thread clockwise around the left index, making a loop that completely encircles the finger. Pinch the intersection of this loop between your right thumb and index finger. Slip your left index finger out of the loop. Now give the loop a half turn away from you by rotating your right hand so that the hand is behind the lucet and the palm faces you. Place the loop on the LH horn and pull the working thread to the right to tighten the loop. The work should now appear as in **Fig. 12**.

N.B.: In the future, this step will be summarized as "Make a Left Twist Loop on the LH horn".

3. Make a Right Twist Loop on the RH horn. Lift the old loop up, off, and over. Pull the working thread to the left to center the knot *but be careful not to tighten it all the way. You will tighten this knot when you lift the next old loop up, off, and over.*

N.B.: In the future, this step will be summarized as "Make a Right Twist Stitch on the RH horn".

4. Make a Left Twist Loop on the LH horn. Lift the old loop up, off, and over. Pull the working thread to the right to center the knot *but be careful not to tighten it all the way. You will tighten this knot when you lift the next old loop up, off, and over.*

N.B.: In the future, this step will be summarized as "Make a Left Twist Stitch on the LH horn".

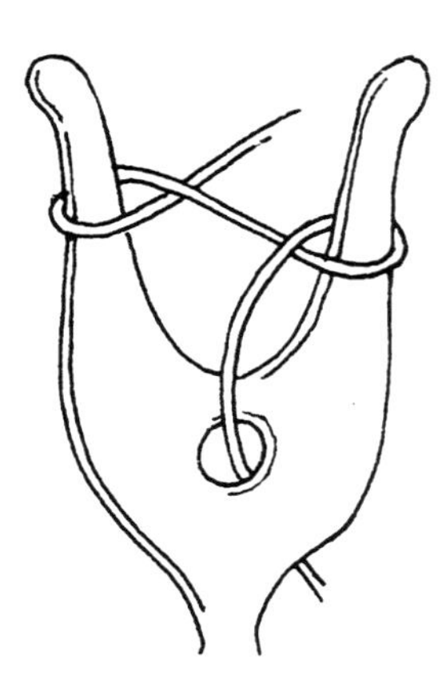

Fig. 12

REPEAT STEPS 3 AND 4.

When lifting a loop, pull first on the back of the loop to tighten the knot, then on the front of the loop to gain enough slack to make the lift. This will both tighten the old knot and give you enough slack to lift the loop. If you need still more slack, pull on the part of the loop that lies IFO the opposite horn.

ENDING OFF THE CORD

When you are finished with this cord, end off as usual for a no-turn cord.

Lay your samples of turned and no-turn twisted stitch cords out before you and take a good look at them. They appear slightly different because of the differing placement of the half hitches on the horns. Now compare the front and back of the no-turn cord. The back will look more "braided" than the front. If you are applying the cord as trim, you may wish to use the back as the "right" side.

A simple pattern can be made by alternating twisted and untwisted stitches. Try making blocks of four or more twisted stitches followed by the same number of untwisted stitches.

MULTIPLE-THREAD CORDS

There are several ways to make cords with more than one thread. We will begin with a single gimp inlay, then work our way up to cords with three working threads. The position of the hands and the lucet remains the same as for the basic cord.

CORD WITH ONE GIMP

A decorative gimp thread floats in and out along the face and back of the braid. The working thread of a lucet cord must be able to withstand a fair degree of abrasion, but a gimp can be quite delicate.

MATERIALS:

Base cord: DMC Cebelia #10
Gimp: Metallic Gold Braid, 1/8 inch or narrower

Method: No Turn

1. Begin a no-turn cord (p. 7, **Single Thread Cord: No-Turn Method,** steps 1 - 4). You should end up with a loop on each horn and two completed knots.

2. Drop a 4-inch tail of gimp down the back of the lucet and hold it with your left hand. Bring the working end of the gimp to the front of the lucet by passing the gimp up and over the completed knots of the cord but under the working thread (**Fig. 13**). With your left thumb, hold the working end of the braid against the front of the lucet.

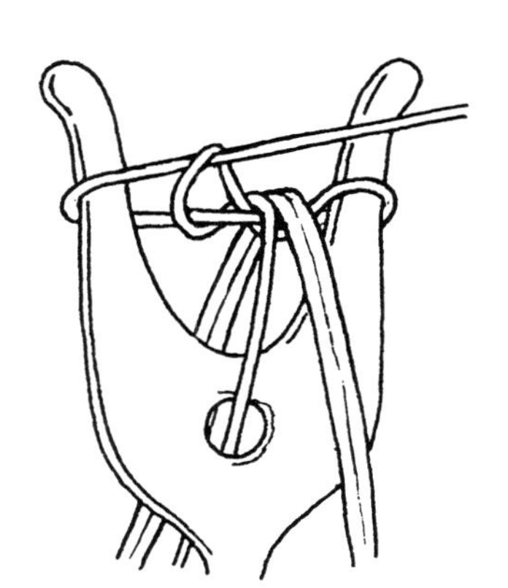

Fig. 13

3. Make a stitch on the RH horn, then a stitch on the LH horn.

4. Now bring the gimp to the *back* of the lucet by passing the gimp up and over the completed knots of the cord but under the working thread of the Cebelia. With your left hand, lightly hold the working end of the gimp against the back of the lucet.

5. With gimp behind the lucet, make a stitch on the RH horn.

6. With gimp behind the lucet, make a stitch on the LH horn. Pass gimp to the front.

7. With gimp IFO the lucet, make a stitch on the RH horn.

8. With gimp IFO the lucet, make a stitch on the LH horn. Pass gimp to the back.

REPEAT STEPS 5 - 8.

ENDING OFF THE CORD

End off as usual for a no-turn cord. Fold the tails of the gimp flat against the back of the work and whip them down with sewing thread.

When you are comfortable handling the gimp in a regular no-turn cord, try adding a gimp thread to a cord with twisted stitches. This is particularly useful for making trims.

The above directions have you float the braid over two stitches on the face of the cord, and two stitches on the back. This produces a cord with tiny spots of gold down the center. Floating the braid over four stitches on the face and four on the back makes the gold much more prominent. You can also float the gimp over six and six, or over four stitches on the face and two on the back, or whatever combination suits you. Softer gimp threads, such as floss or chenille, can also be used. You can even thread beads on a gimp. All in all, this is a simple but versatile technique for making decorative cords.

CORD WITH TWO GIMPS

MATERIALS:

Base cord: DMC Coton Perle #3 (DK)
Gimps: Two colors DMC Coton Perle #3 (one LT, one MED)

Method: No Turn

1. With DK, begin a no-turn cord (steps 1 - 4, **Single Thread Cord: No-Turn Method**). You should end up with a loop on each horn of the lucet and two completed knots in the middle.

2. Drop a 4-5 inch tail of LT down the back of the lucet. Bring the working end of the LT thread to the front of the lucet as you did in step 2 of the Cord with One Gimp.

3. With DK, make a stitch on the RH horn.

4. With DK, make a stitch on the LH horn.

5. Now bring the LT gimp to the back of the lucet. Drop a 4-5 inch tail of MED down the back of the lucet. Bring the working end of the MED thread to the front of the lucet (**Fig. 14**).

6. With DK, make a stitch on the RH horn.

7. With DK, make a stitch on the LH horn.

8. Bring the MED gimp to the back of the lucet and the LT gimp to the front.

9. With DK, make a stitch on the RH horn.

10. With DK, make a stitch on the LH horn.

11. Bring the LT gimp to the back of the lucet and the MED gimp to the front.

REPEAT STEPS 6 - 11.

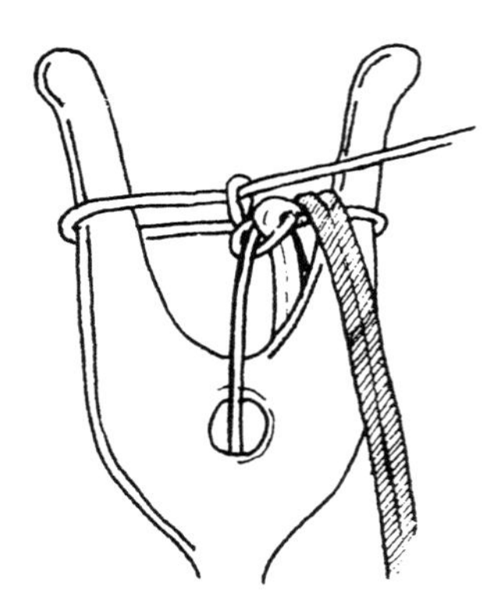

Fig. 14

ENDING OFF THE CORD

Leave the gimp that is already at the back of the cord where it is. Pass the other gimp to the back as well. Make a stitch on the RH horn and a stitch on the LH horn. End off as usual for a no-turn cord. Weave the tails of the gimps into the floats along the back of the work.

This cord is especially pretty worked in size 5 pearl cotton with twisted stitches.

CORDS WITH THREAD EXCHANGE

Exchanging the gimp and working thread is one way to make a multi-color cord.

MATERIALS:

Base cord and Gimp: Two colors of DMC Coton Perle #3, one LT and one DK

Method: No Turn

1. With DK, make a stitch on the RH horn and a stitch on the LH horn. Drop a tail of LT down the back of the lucet. Bring the working end of the LT thread to the front of the lucet. Make a stitch of DK on the RH horn. Make a stitch of DK on the LH horn (the LT thread is floating on the face of the cord). Now pass the gimp (LT) to the back. With DK make a stitch on the RH horn and a stitch on the LH horn.

2. Pass the gimp (LT) to the front. With DK make a stitch on the RH horn and stitch on the LH horn.

3. Pass the gimp (LT) to the back. With DK make a stitch on the RH horn and stitch on the LH horn. You are making a DK base cord with tiny LT spots.

REPEAT STEPS 2 AND 3 UNTIL YOU WISH TO CHANGE THE COLOR OF THE BASE CORD. END WITH STEP 3 (GIMP IN BACK).

4. LT FOR DK EXCHANGE: Drop the DK thread behind the horns and let it hang there. Bring the LT thread IFO the RH horn. Make a LT stitch on the RH horn of the lucet. Pull on the DK thread to tighten the last DK knot. Now make a LT stitch on the LH horn. Again, pull on the DK thread to tighten the last DK knot. The work should appear as in **Fig. 15**.

5. Pass the DK thread (now functioning as the gimp) to the front of the lucet. With LT, make a stitch on the RH horn and a stitch on the LH horn.

6. Pass the gimp (DK) to the back. With LT make a stitch on the RH horn and stitch on the LH horn. Now you are making a LT base cord with tiny DK spots.

REPEAT STEPS 5 AND 6 UNTIL YOU WISH TO CHANGE THE COLOR OF THE BASE CORD. END WITH STEP 6 (GIMP IN BACK).

7. DK FOR LT EXCHANGE: Drop the LT thread behind the horns of the lucet and let it hang there. Bring the DK thread IFO the RH horn. Make a DK stitch on the RH horn of the lucet. Pull on the LT thread to tighten the last LT knot. Now make a DK stitch on the LH horn. Again, pull on the LT thread to tighten the last LT knot. The work should appear as in **Fig. 16**.

REPEAT STEPS 2 - 7.

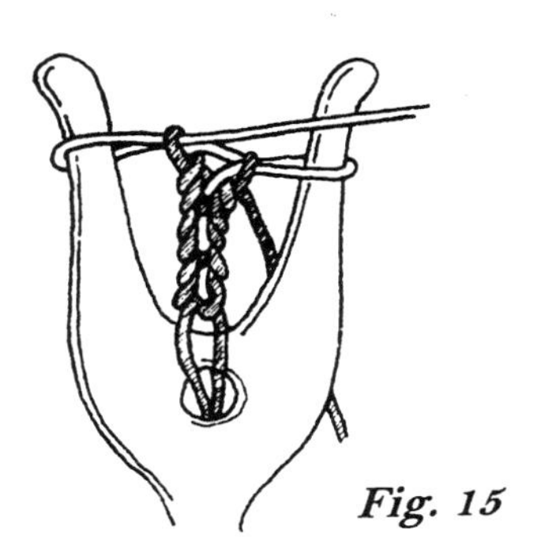

Fig. 15

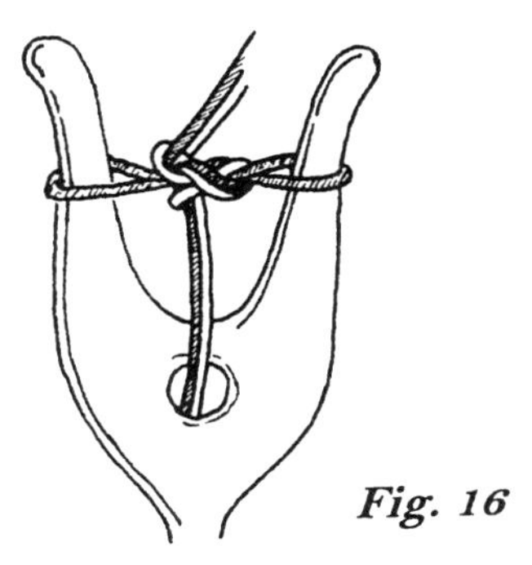

Fig. 16

ENDING OFF THE CORD

When you are finished with this cord, end off as usual for a no-turn cord. Darn the tails of the gimp into the floats along the back of the work.

This cord is particularly nice when made with twisted stitches.

THREE-COLOR CORD WITH THREAD EXCHANGE

MATERIALS:

Base cord and Gimps: Three colors of DMC Coton Perle #3 (LT, MED, and DK)

Method: No Turn

1. With DK, make a stitch on the RH horn and a stitch on the LH horn. Drop a tail of LT down the back of the lucet and bring the working end of the LT thread to the front of the lucet. With DK make a stitch on the RH horn and a stitch on the LH horn.

2. Pass the LT to the back. Drop a tail of MED down the back of the lucet. Bring the working end of the MED thread to the front of the lucet. With DK make a stitch on the RH horn and a stitch on the LH horn.

3. Pass the MED to the back and the LT to the front. With DK make a stitch on the RH horn and a stitch on the LH horn.

4. Pass the LT to the back and the MED to the front. With DK make a stitch on the RH horn and a stitch on the LH horn.

REPEAT STEPS 3 AND 4 UNTIL YOU WISH TO CHANGE THE COLOR OF THE BASE CORD. END WITH STEP 4 (LT IN BACK).

5. MED FOR DK EXCHANGE: Drop the DK thread behind the horns and let it hang there. Pass the LT thread to the front of the lucet. With MED make a stitch on the RH horn of the lucet. Pull on the DK thread to tighten the last DK knot. Now make a MED stitch on the LH horn. Again, pull on the DK thread to tighten the last DK knot.

6. Pass the LT to the back and the DK to the front. With MED make a stitch on the RH horn and a stitch on the LH horn.

7. Pass the DK to the back and the LT to the front. With MED make a stitch on the RH horn and a stitch on the LH horn.

REPEAT STEPS 6 AND 7 UNTIL YOU WISH TO CHANGE THE COLOR OF THE BASE CORD. END WITH STEP 7 (LT IN FRONT).

8. LT FOR MED EXCHANGE: Drop the MED thread behind the horns and let it hang there. Pass the DK thread to the front of the lucet. With LT make a stitch on the RH horn of the lucet. Pull on the MED thread to tighten the last MED knot. Now make a LT stitch on the LH horn. Again, pull on the MED thread to tighten the last MED knot.

9. Pass the DK to the back and the MED to the front. With LT make a stitch on the RH horn and a stitch on the LH horn.

10. Pass the MED to the back and the DK to the front. With LT make a stitch on the RH horn and a stitch on the LH horn.

REPEAT STEPS 9 AND 10 UNTIL YOU WISH TO CHANGE THE COLOR OF THE BASE CORD. END WITH STEP 10 (DK IN FRONT).

11. DK FOR LT EXCHANGE: Drop the LT thread behind the horns and let it hang there. Pass the MED thread to the front of the lucet. With DK make a stitch on the RH horn of the lucet. Pull on the LT thread to tighten the last LT knot. Now make a DK stitch on the LH horn. Again, pull on the LT thread to tighten the last LT knot.

REPEAT STEPS 3 - 11

ENDING OFF THE CORD

Leave the gimp that is already at the back of the cord where it is. Pass the other gimp to the back. With the current working thread, make a stitch on the RH horn and a stitch on the LH horn. End off as usual for a no-turn cord, and then weave the tails of the gimps into the floats along the back of the work. Or braid the three tails together, or use them to attach a tassel, or find some whimsical finish that I haven't thought of!

CORDS WITH TWO & THREE WORKING THREADS

Adding a second working thread makes a chevron pattern along the length of the cord. A second (or even third) thread can also be made into picots, something which is not possible with only one working thread.

TURNED CORD WITH TWO WORKING THREADS: Method 1

In this method, different color loops are stacked on the horns and lifted off one at a time. Although this cord can be worked with the threads carried in the hand, using two kumi himo bobbins greatly improves the evenness of the tension.

1. Select a LT and a DK thread and wind each onto a kumi himo bobbin. Holding the two threads together as one, drop a six-inch tail from front to back through the hole of the lucet.

2. Keeping the LT thread below the DK one, wrap the thread around the horns of the lucet in the usual figure 8. Now bring the LT thread IFO the RH horn and *above the first wrap*. The DK thread hangs behind the LH horn. **Fig. 17.**

3. Pick up *only* the lower LT wrap on the RH horn and lift it up past *both* the DK wrap *and* the upper LT thread, then off the tip of the RH horn as usual. Drop the wrap as usual behind the RH horn. Pull the LT thread to the right to tighten the loop on the horn The work should appear as in **Fig. 18** (p. 16).

4. Leaving the LT thread behind the horn, bring the DK working thread IFO the RH horn. Now pick up *only* the lower DK wrap on the RH horn and lift it up past *both* the LT loop *and* the upper DK thread, then off the tip of the RH horn as usual. Drop the wrap as usual behind the RH horn. Pull the DK thread to the right to tighten the loop on the horn. The work should appear as in **Fig. 19** (p. 16).

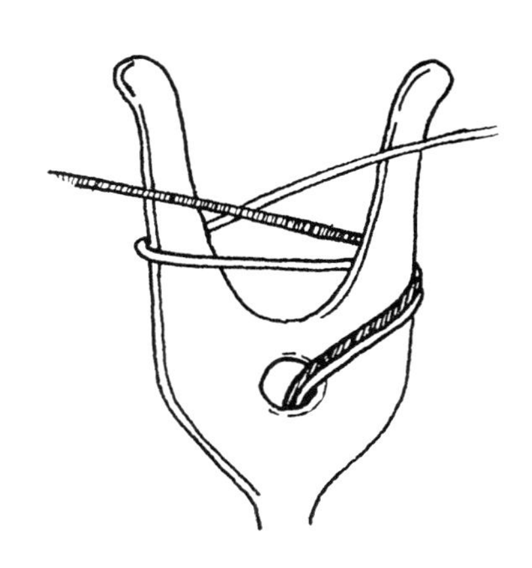

Fig. 17

5. Turn the lucet from right to left.

6. Leaving the LT thread IFO the RH horn, pass the DK thread to the back of the lucet. Now pick up *only* the lower LT loop on the RH horn and lift it up past *both* the DK loop *and* the upper LT thread, then off the tip of the horn as usual. Drop the loop as usual behind the RH horn. Pull the LT thread to the right to tighten the loop on the horn.

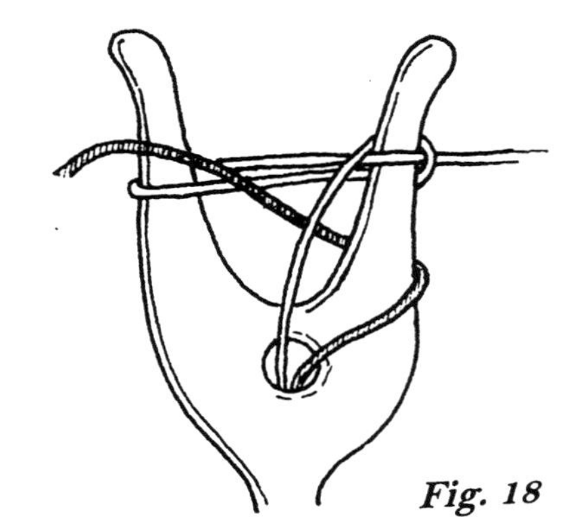

Fig. 18

7. Leaving the LT thread behind the horn, bring the DK working thread IFO the RH horn. Now pick up *only* the lower DK loop on the RH horn and lift it up past *both* the LT loop *and* the upper DK thread, then off the tip of the horn as usual. Drop the loop as usual behind the RH horn. Pull the DK thread to the right to tighten the loop on the horn.

REPEAT STEPS 5 - 7.

You may find it helpful to remember that after you have turned the lucet, the *l*ower *l*oop should be *l*ight. Keep the loops on the horns neat and in order. Do not lay the work aside in the middle of a repeat or you may get hopelessly lost. Always complete step 7 before ending your cordmaking session.

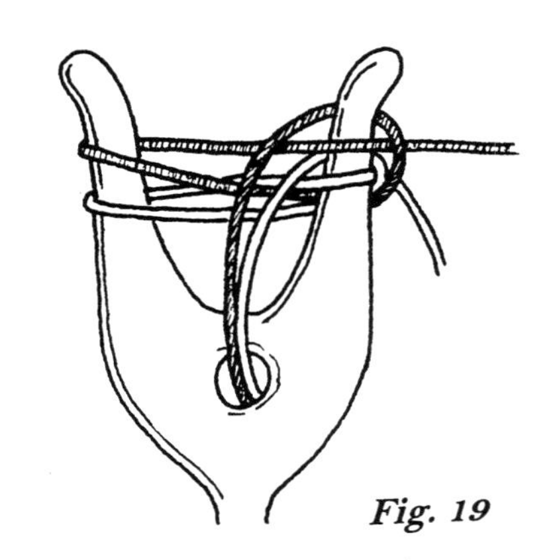

Fig. 19

ENDING OFF THE CORD

This is much the same as ending off the basic single thread cord. The two threads are held together as at the beginning of the cord.

1. Turn the lucet as though you were going to start a new stitch, but instead of lifting the loop, cut the threads, leaving about a six-inch tail.
2. Thread the tail down through both loops on the RH horn. Slip the loops off the horn together. Pull on the tail until the loops close.
3. Now thread the tail down through both loops on the LH horn. Slip the loops off the horn together. Pull on the tail until the loops close.

TURNED CORD WITH TWO WORKING THREADS: Method 2

This method carries one loop on the horns but produces a two-color cord. The tension is difficult to keep even with the threads carried in the hand. You will need to wind the individual threads onto kumi himo bobbins. This structure allows you to make picots.

1. Select a LT and a DK thread and wind each onto a kumi himo bobbin. Holding the two threads together as one, drop a six-inch tail from front to back through the hole of the lucet. Wrap the lucet as in step 2 of the Turned Cord with Two Working Threads, Method I.

2. Pick up *both* of the threads in the lower wrap on the RH horn. Treating these two threads as one, lift them up past the upper LT wrap, then off and over the horn. Gently pull the LT thread to the left to center the knot. The work should appear as in **Fig. 20.**

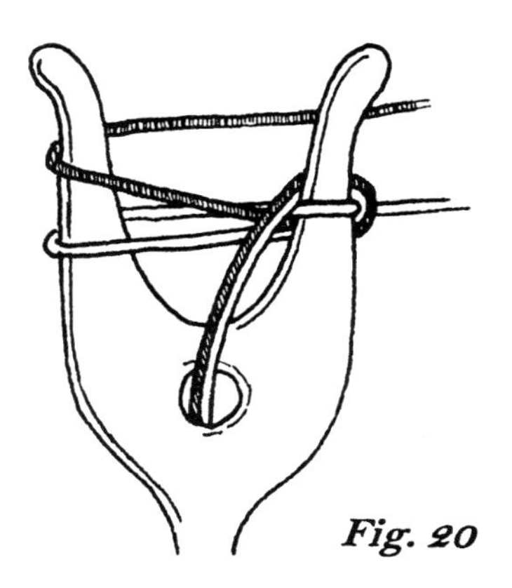

Fig. 20

3. Keeping the LT thread in back of the lucet, place the DK thread IFO the RH horn. Lift the LT loop up past the DK thread, then off and over. Pull the LT thread back and away from you to tighten the knot. You should have a single, DK loop on the RH horn and a double, LT and DK loop on the LH horn (**Fig. 21**).

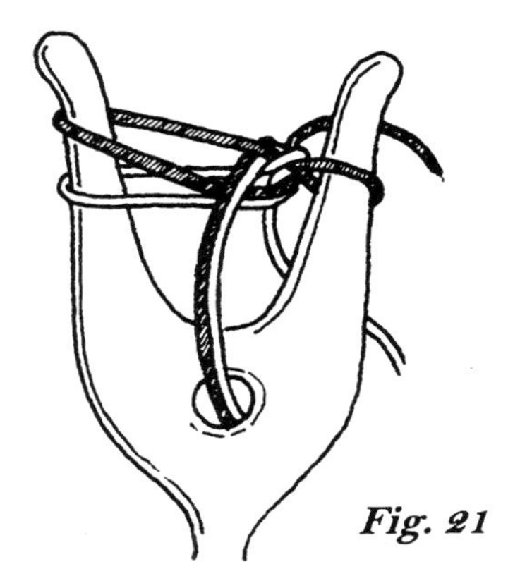

Fig. 21

4. Turn the lucet from right to left.

5. Place the DK thread behind the RH horn and the LT thread IFO the RH horn. Pick up *both* of the threads in the lower wrap on the RH horn. Treating these two threads as one, lift them up past the upper (LT) wrap, then off and over the horn. Gently pull the LT thread to the left to center the LT knot. Now pull the DK thread to the right to center the DK knot. You may need to first pull on the back of the loop on the LH horn. The work should appear as in **Fig. 22**, with a LT loop on the RH horn and a DK loop on the LH horn.

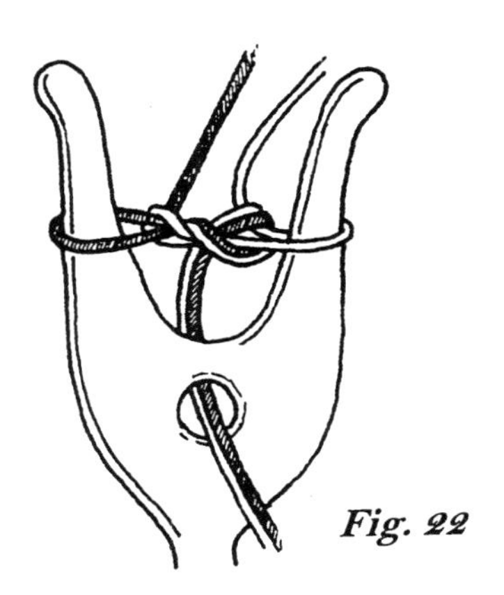

Fig. 22

6. Keeping the LT thread in back of the lucet, place the DK thread IFO the RH horn. Lift the LT loop up past the DK thread, then off and over the horn. Pull the LT thread back and away from you to tighten the knot. You should have a DK loop on each horn.

7. Turn the lucet from right to left.

8. Keeping the LT thread IFO the RH horn, place the DK thread behind the horn. Pick up the lower (DK) loop on the RH horn. Lift the loop up past the upper LT wrap, then off and over the horn. Gently pull the DK thread to the right to tighten the knot. The work should appear as in **Fig. 23**, with a LT loop on the RH horn and a DK loop on the LH horn.

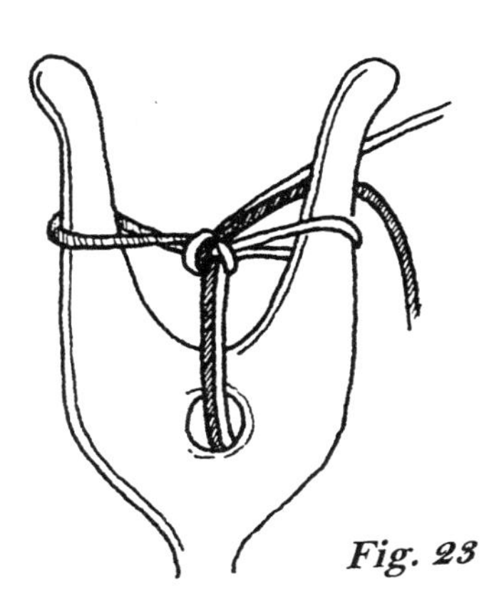

Fig. 23

9. Keeping the LT thread in back of the lucet, place the DK thread IFO the RH horn. Lift the LT loop up past the DK thread, then off and over the horn. Pull the LT thread back and away from you to tighten the knot. You should have a DK loop on each horn.

REPEAT STEPS 7 - 9.

ENDING OFF THE CORD

End as you would a Turned Cord With Two Working Threads, Method I

PICOTS

Now for those picots I've been promising you. Work until you have completed step 7. To make a picot, substitute this modified step 8:

8. Keeping the LT thread IFO the RH horn, place the DK thread behind the horn. Pick up the lower (DK) loop on the RH horn. Lift the loop up past the upper LT wrap, then off and over the horn as usual. Gently pull the DK thread to the right to tighten the knot. *Place the LT thread IFO the horn again and lift the LT loop up, off, and over. Pull the LT thread to the left to tighten the knot. Repeat from * until your picot is big enough. The work should appear as in **Fig. 24.** Now proceed on to step 9, which will close the picot for you.

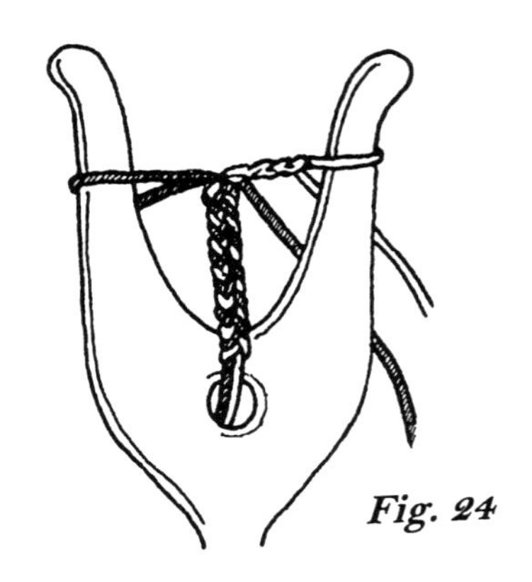

Fig. 24

Upon completion of step 9, your picot will look a little loose. Don't worry about it; just turn the lucet (step 7) and go on to the *unmodified* step 8. Pulling on the DK thread to tighten the knot in step 8 will further tighten the picot.

Make sure you have at least two plain stitches between picots. This preserves the structural integrity of the cord.

It is, of course, possible to make this cord with twisted stitches. To set up the lucet, hold the LT and DK together as one. Make an initial Right Twist Loop on the RH horn, then turn the lucet and make another Right Twist Loop on what is now the RH horn (**Fig. 25**). *Turn the lucet, make a Right Twist Stitch on the RH horn in LT, then make a Right Twist Stitch on the RH horn in DK, and repeat from *.

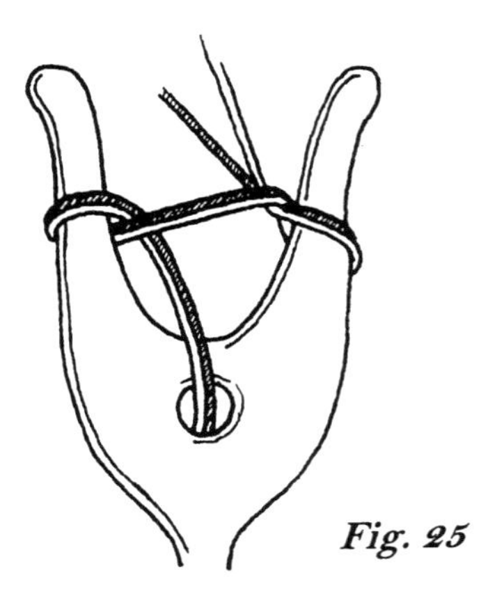

Fig. 25

TURNED CORD WITH TWISTED STITCHES AND TWO WORKING THREADS

1. Holding the two threads together as one, drop a six-inch tail from front to back through the hole of the lucet. Pass both working threads to the back of the lucet. With the LT thread, make a Right Twist Loop on the RH horn. Above the LT loop, make a Right Twist Loop with DK thread.

2. Turn the lucet. Make a LT Right Twist Loop on what is now the RH horn. Above this LT loop, make a Right Twist Loop with DK thread. The work should appear as in **Fig. 26.**

3. Turn the lucet again. Position both threads behind the RH horn.

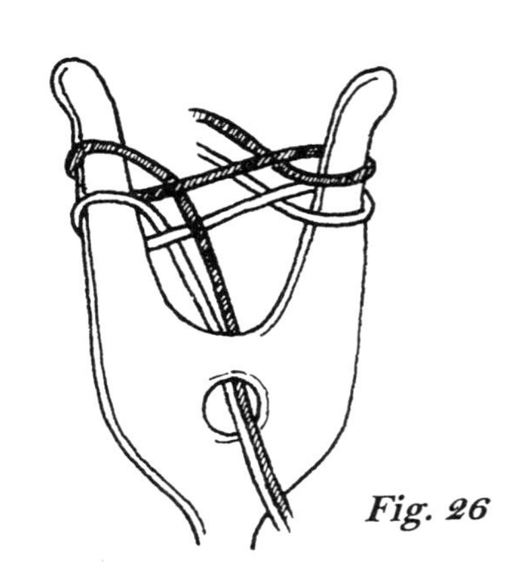

Fig. 26

4. With the LT thread, make a Right Twist Loop on what is now the RH horn (**Fig. 27**). Lift the lower LT loop up past both the DK loop and the upper LT loop, then off over the tip of the horn. Drop the loop as usual behind the horn. Gently pull the LT working thread to the right to tighten the new loop on the horn.

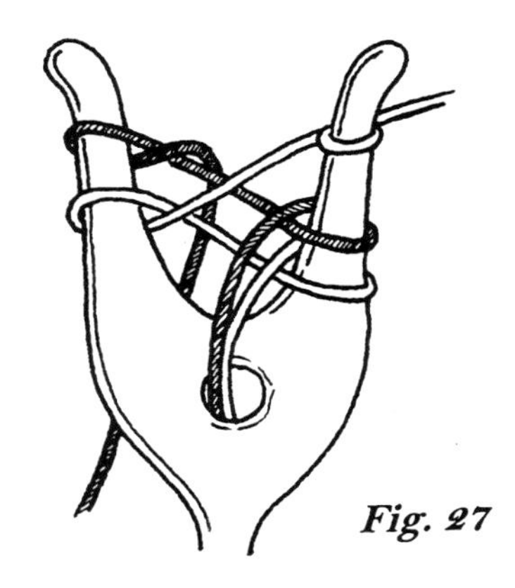
Fig. 27

5. With the DK thread, make a Right Twist Loop on the RH horn (**Fig. 28**). Lift the lower DK loop up past both the LT loop and the upper DK loop, then off over the tip of the horn. Drop the loop as usual behind the horn. Gently pull the DK working thread to the right to tighten the new loop on the horn. Now pull both working threads to the left to center the knots just made.

REPEAT STEPS 3 - 5.

As with all twisted-stitch cords, you must leave the old knot a little loose at first. This gives you enough slack to manipulate the loop. Remember to tighten the old knot by pulling on the back of the loop immediately before you lift the loop up, off, and over.

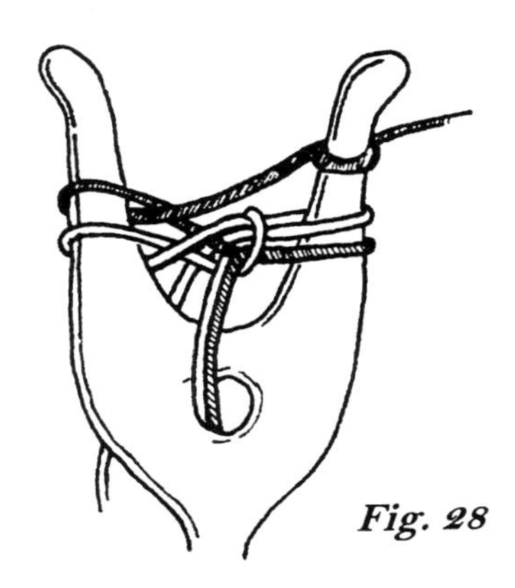
Fig. 28

ENDING OFF THE CORD

End this cord as you did the preceding one. You may need to close the last knots by pulling on the backs of the RH loops before you thread the tail down through the loops.
Fopllowing, you will find no-turn versions of the above two cords, suitable for experiments with gimps and thread exchange.

NO-TURN CORD WITH TWO WORKING THREADS

1. Wind the LT and DK threads and set up the lucet as on steps 1 and 2 of the Turned Cord with Two Working Threads, Method I (p. 15).

2. Pick up *only* the LT thread wrap on the RH horn and lift it up past both the DK wrap and the upper LT thread, then off and over the tip of the horn. Pull the LT working thread to the right to tighten the loop. Now bring the DK thread IFO the RH horn. Lift the lower DK wrap up past both the LT loop and the upper DK thread. Tighten as usual.

3. Without catching the DK thread, wrap the LT thread counterclockwise around the LH horn. Drop the bobbin to the back of the lucet between the horns. Lift *only* the LT loop up, off, and over. Center the knot. Now pass the DK working thread *underneath* the hanging LT thread. Wrap the DK thread counterclockwise around the LH horn. Drop the DK bobbin to the back. Lift the DK loop. Center the knot.

4. Without catching the DK thread, lay the LT thread across and IFO the RH horn, lift the LT loop, and tighten. Without catching the DK thread, lay the DK thread across and IFO the RH horn, lift the DK loop, and tighten.

REPEAT STEPS 3 - 4, ENDING THE CORD WITH STEP 3.

If, in step 3, of the NO-TURN CORD WITH TWO WORKING THREADS, you pass the DK working thread *over* the hanging LT thread, you will get a different pattern. Whichever method you choose, be consistent.

NO-TURN CORD WITH TWISTED STITCHES AND TWO WORKING THREADS

1. Thread the LT and DK tails through the hole. Pass the working threads to the back. Make a LT Right Twist Loop on the RH horn then a DK Right Twist Loop above the LT one.

2. Make a LT Left Twist Loop on the LH horn, then a DK Left Twist Loop above the LT one (**Fig. 29**).

3. Make a LT Right Twist Loop on the RH horn. Lift the lower LT loop up and off past both the DK loop and the upper LT loop. Pull the working thread to the left to center the knot.

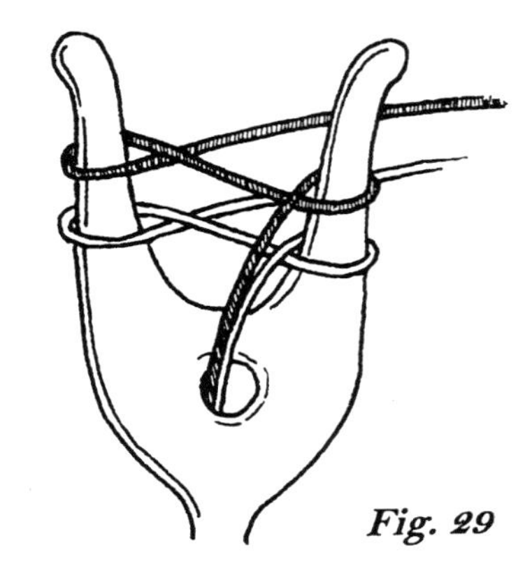

Fig. 29

4. Pass the DK working thread *underneath* the hanging LT thread. Make a DK Right Twist Loop on the RH horn. Lift the lower DK loop up and off past both the LT loop and the upper DK loop. Pull the working thread to the left to center the knot.
5. Without catching the DK thread, make a LT Left Twist Loop on the LH horn. Lift the lower LT loop up and off past both the DK loop and the upper LT loop. Pull the working thread to the right to center the new knot.
6. Pass the DK working thread *underneath* the hanging LT thread. Make a DK Left Twist Loop on the LH horn. Lift the lower DK loop up past both the LT and upper DK loop, then off and over as usual. Pull the working thread to the right to center the new knot.

REPEAT STEPS 3 - 6.

ENDING OFF THE NO-TURN CORDS

Hold the two threads together as at the beginning of the cord.
1. Cut the threads, leaving about a six-inch tail.
2. Thread the tail down through both loops on the RH horn. Slip the loops off the horn together. Pull on the thread until the loops close.
3. Now thread the tail down through both loops on the LH horn. Slip the loops off the horn together. Pull on the tail until the loops close.

TURNED CORD WITH THREE WORKING THREADS

This cord works up more easily with the threads wound on kumi himo bobbins.

1. Select a LT, MED and DK thread and wind each onto a kumi himo bobbin. Holding the threads together as one, drop them F to B through the hole of the lucet. Wrap the lucet in the normal figure-8. Leave the LT thread IFO the RH horn. Shift the MED and DK threads to the back of the horn (**Fig. 30**).

Be careful throughout this cord to keep the threads in their proper order on the horns.

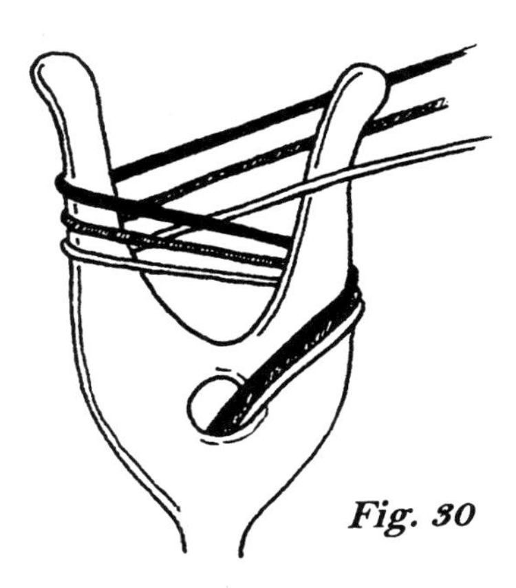

Fig. 30

2. Pick up only the LT lower wrap and lift it up, off, and over. Pull the LT thread to the right to tighten the loop on the horn (**Fig. 31**).

3. Place the MED thread IFO the RH horn. Now pick up only the MED lower wrap and lift it up, off, and over. Pull the MED thread to the right to tighten the loop on the horn.

4. Place the DK thread IFO the RH horn. Pick up only the DK lower wrap and lift it up, off, and over. Pull the DK thread to the right to tighten the loop on the horn. The work should appear as in **Fig. 32**.

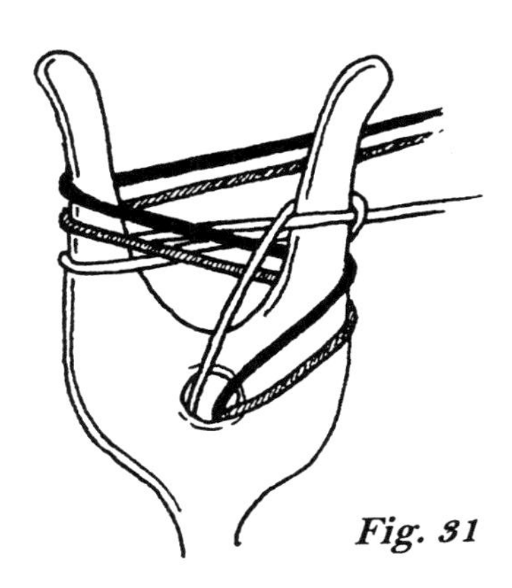

Fig. 31

5. Turn the lucet. Leaving the LT thread IFO the RH horn, shift the MED and DK threads to the back of the horn. Repeat steps 2 - 4 on what is now the RH horn.

6. Turn the lucet. If necessary, pull all three working threads to the right to center the knots.

7. Leaving the LT thread IFO the RH horn, shift the MED and DK threads to the back of the horn. Pick up only the LT loop and lift it up, off, and over. Pull the LT working thread to the right to tighten the loop on the horn.

8. Place the MED thread IFO the RH horn. Now pick up only the MED loop and lift it up, off, and over. Pull the MED thread to the right to tighten the loop on the horn.

9. Now place the DK thread IFO the RH horn. Pick up only the DK loop and lift it up, off, and over. Pull the DK thread to the right to tighten the loop on the horn. Turn the lucet and center the knots.

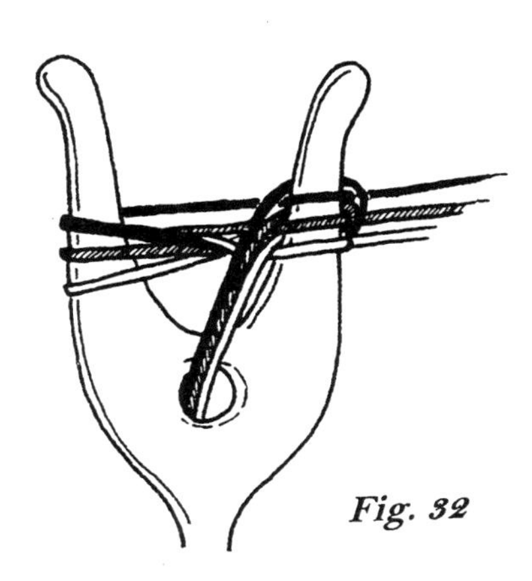

Fig. 32

REPEAT STEPS 7 - 9.

This same LT-MED-DK stitch sequence can be applied to a no-turn cord. To make a cord with twisted stitches, set up the lucet as in **Fig. 33**. Make a Right Twist Loop in LT on the RH horn, and lift the lower LT loop up, off, and over. Follow the same procedure with first the MED thread and then the DK, then turn the lucet. This is a very flat cord.

All of the cords in this section can incorporate gimps and thread exchange. Experiment and have fun!

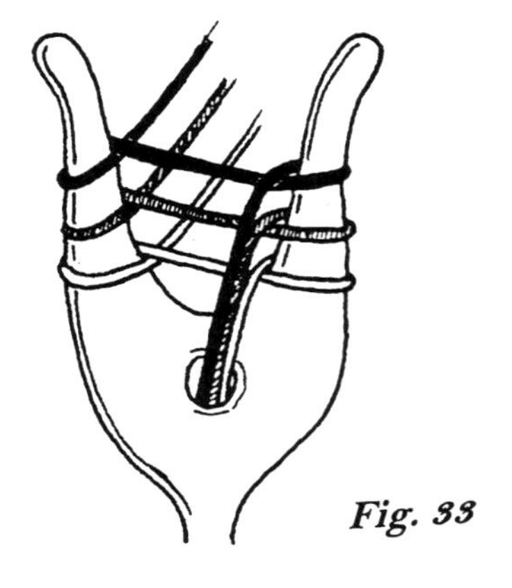

Fig. 33

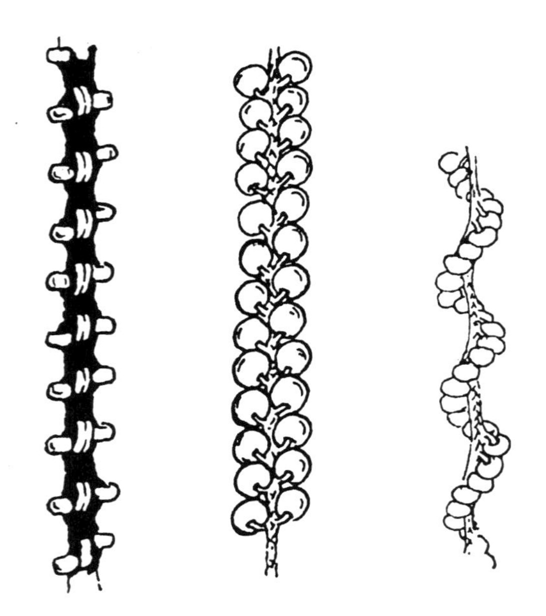

Some beaded cords. See pp. 24 - 25.

BEADED CORDS

MATERIALS:

DMC Coton Perle #8; #11 seed beads

Method: No Turn

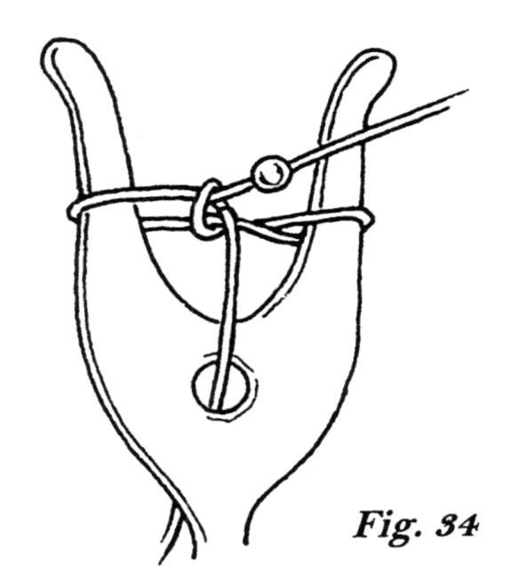

Fig. 34

1. String the beads onto the working thread and push them out of the way, close to the ball. Wrap the lucet in the usual figure 8.

2. Begin the cord by making a stitch on the RH horn, then a stitch on the LH horn.

3. Slide a bead up the thread and IFO the RH horn. The bead should rest above the loop, between the knots and the inner edge of the horn. Wrap the thread counterclockwise around the RH horn (**Fig. 34**).

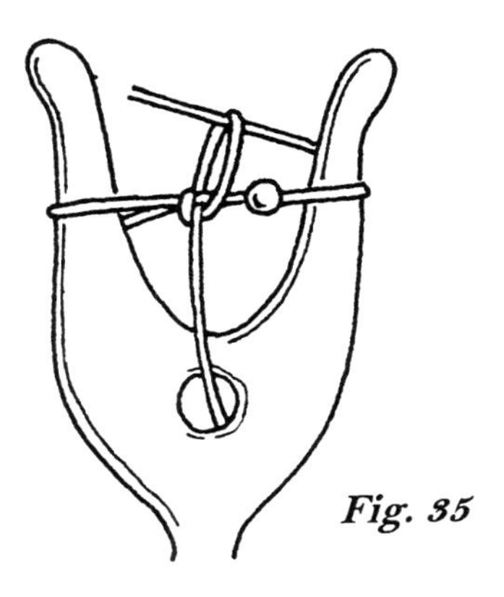

Fig. 35

4. Lift the loop off the RH horn *and over both the horn and the bead.* The bead should be able to move freely along the new RH loop instead of being caught against the cord (**Fig. 35**). Center knot.

5. Make a stitch on the LH horn.

6. Slide the bead on the RH loop up against the last knot and hold it there with your left thumbnail. *Keeping hold of the bead,* make a stitch on the RH horn in the usual manner.

7. Still holding the bead, make a stitch on the LH horn.

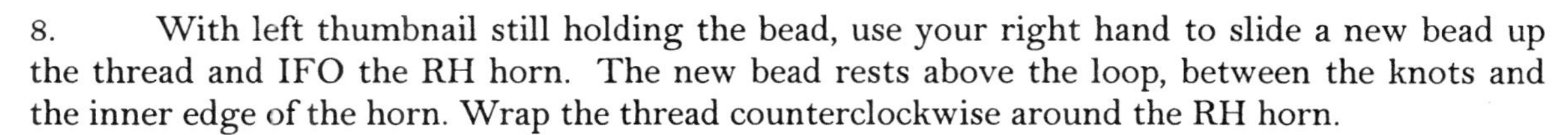

8. With left thumbnail still holding the bead, use your right hand to slide a new bead up the thread and IFO the RH horn. The new bead rests above the loop, between the knots and the inner edge of the horn. Wrap the thread counterclockwise around the RH horn.

N.B.: Form now on, this step will be summarized as "Drop a bead IFO the RH horn".

9. Still holding old the bead, lift the loop off the RH horn *and over both the horn and the bead.* The bead should be able to move freely along the new RH loop instead of being caught against the cord. Center knot.

N.B.: Form now on, this step will be summarized as "Make a beaded stitch on the RH horn".

10. *Still* holding old the bead, make a stitch on the LH horn. Now you can finally let go of that old bead!

11. Slide the bead on the RH loop up against the knot and - you guessed it - *hold it there with your left thumbnail.* Hang on to the bead and make a stitch on the RH horn in the usual manner.

12. Still holding the bead, make a stitch on the LH horn.

REPEAT STEPS 8 - 12

ENDING OFF THE CORD

End off as usual for a no-turn cord.

In spite of all the above verbiage, this is really a simple, four-step operation:

A) Drop a bead IFO the RH horn and make a beaded stitch on the RH horn.

B) Make a stitch on the LH horn.

C) Slide bead to knot and hold it; make a stitch on the RH horn.

D) Make a stitch on the LH horn.

The keys to a neatly beaded cord are to get the loop *over the bead* as well as the horn, and to hang on to that old bead until you're ready to slide the next bead to the knot.

The structure of the above cord alternates one beaded and one unbeaded stitch along the RH edge, leaving the LH edge unbeaded. If you leave the LH edge plain, but bead every stitch on the RH edge, the closer spacing of the beads will force the cord into a curve. If continued long enough, the curved cord begins to corkscrew. You can make some interesting dangles this way, or a three-dimensional, coiled trim which can be applied to the edge of a hem.

You can also space the beads more widely. Try beading every third or fourth stitch along the RH edge.

Beading the LH edge requires a slightly different technique:

1. Begin cord with a stitch on the RH horn and a stitch on the LH horn, then another stitch on the RH horn.

2. Slide a bead up the working thread and *behind* the LH horn. The bead should rest above the loop, between the knots and the inner edge of the horn. Wrap the thread counterclockwise around the LH horn. **Fig. 36**. [I.e., drop a bead behind the LH horn].

3. Lift the LH loop off and over both the horn *and the bead.* Center knot. [I.e., make a beaded stitch on the LH horn].

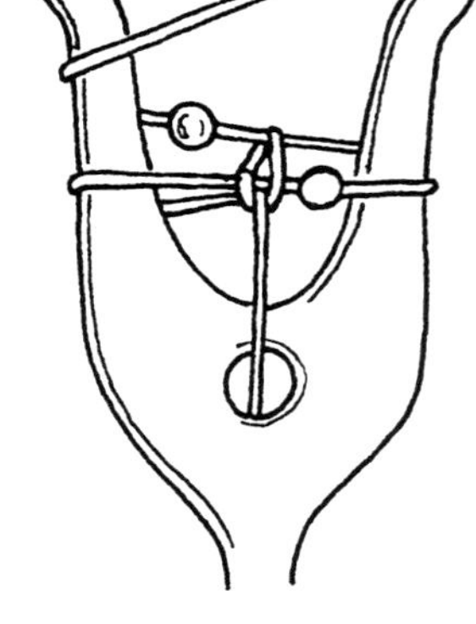

Fig. 36

4. Make a stitch on the RH horn.

5. Slide the bead around to the front of the LH horn and hold it against the knots with your left thumbnail. Make a stitch on the LH horn.

6. Make a stitch on the RH horn.

REPEAT STEPS 2 - 6

Now try a cord with beading along both edges:

1. Begin cord with a stitch on the RH horn and a stitch on the LH horn.

2. Drop a bead IFO the RH horn and make a beaded stitch on the RH horn.

3. Make a stitch on the LH horn.

4. Slide RH bead to knot and hold it; make a stitch on the RH horn.

5. Drop a bead behind the LH horn and make a beaded stitch on the LH horn.

6. Make a stitch on the RH horn.

7. Slide LH bead IFO LH horn and hold it; make a stitch on the LH horn.

REPEAT STEPS 2 - 7.

These cords can all be made with twisted stitches. Just make sure that the bead is properly placed when you drop it, and that the loop in the beaded stitch gets lifted over the bead as well as the horn. You can also add gimps, or bead cords with multiple working threads. Combining different techniques in the same cord is fun and lets you design your own PASSEMENTERIE.

BEADED CORD WITH TWO GIMPS

This is an example of the sort of fancy cord that results from combining several techniques and materials. This particular cord makes a nice bracelet or choker. Substitute 4 mm pearls for the beads and narrow gold braid for the chenille, and you have a stunning trim for a bodice or doublet.

MATERIALS:

Base cord: DMC Cebelia #10; 4 mm beads; Gimp A: Chenille Bayberry yarn;

Gimp B: DMC Coton Perle #8; #11 seed beads

Method: No-Turn

1. String the 4 mm beads onto the Cebelia. String the #11 seed beads onto gimp B. Push the beads out of the way, close to their respective balls. With the Cebelia, wrap the lucet in the usual figure 8.
2. Begin cord in Cebelia with a stitch on the RH horn and a stitch on the LH horn.
3. Drop a 4 - 5 inch tail of Gimp A down the back of the cord and hold it there. Pass the working end of Gimp A to the front of the lucet.
4. Drop a bead IFO the RH horn. Make a beaded stitch on the RH horn.
5. Make a stitch on the LH horn.
6. Slide bead to knot and hold it. Make a stitch on the RH horn.
7. Make a stitch on the LH horn.
8. Pass Gimp A to the back of the lucet.
9. Drop a 4 - 5 inch tail of Gimp B down the back of the cord and hold it there. Pass the working end of Gimp B to the front of the lucet. Use your left thumb and index finger in a pincer grip to hold the bead and the gimps in their proper places.
10. Make a stitch on the RH horn.
11. Drop a bead behind the LH horn. Make a beaded stitch on the LH horn.
12. Make a stitch on the RH horn.
13. Slide bead IFO LH horn and hold against knot; make a stitch on the LH horn.
14. Slide a bead up Gimp B. With your left thumbnail hold the bead against the face of the cord. Now pass gimp B to the *back* of the lucet and Gimp A to the front. *Make sure the bead stays on the face of the cord.* Use your left thumb and index finger to hold the beads on the face of the cord and the gimps in their proper places.
15. Drop a bead IFO the RH horn. Make a beaded stitch on the RH horn.
16. Make a stitch on the LH horn.
17. Slide bead to knot and hold it; make a stitch on the RH horn.
18. Make a stitch on the LH horn.
19. Pass Gimp A to the back of the lucet, and Gimp B to the front.

REPEAT STEPS 10 - 19

LEFT-HANDED LUCET

LEFT-HANDED SINGLE-THREAD CORD

The right hand holds the lucet, horns away from you. The left manipulates the thread.

1. Tread a six-inch tail of crochet cotton from F to B through the hole of the lucet. With your right hand, hold the tail against the back of the lucet. Wrap the thread from the ball twice around the little finger of your left hand, then pass the thread across the palm. Hold the thread between your left thumb and index finger.

2. Wrap the thread around the horns of the lucet in a figure 8. Do this by taking the thread clockwise around to the back of the LH horn, bringing the thread forward between the horns, then wrapping the thread counterclockwise around the RH horn. Now bring the thread IFO the LH horn and *above the first wrap*. **Fig. 37.**

3. Slip your left thumb and index under the working thread and behind the LH horn. Grasp the back of the lower of the two threads lying between the horns. Pull this thread counterclockwise around to the front of the LH horn and without twisting the thread lift it up past the upper thread, off over the tip of the horn, and to the back of the lucet **(Fig. 38)**. *The upper thread must remain in its place IFO the LH horn.* Drop the just-lifted thread behind the LH horn. You have a loop encircling each horn and a loose knot sitting behind the LH horn. Pull the working thread to the left to tighten the loop against the horn **(Fig. 39)**.

4. Using your right hand, turn the lucet over from left to right. The work should now appear as in **Fig. 40**, *with the working thread lying above the loop on the LH horn.*

5. Slip your left thumb and index finger under the working thread and behind the LH horn. Grasp the back of the loop on the LH horn. Pull this part of the loop counterclockwise around to the front of the LH horn. Without twisting the loop lift it up past the upper thread, off over the tip of horn, and to the back of the lucet (this is the same motion shown in **Fig. 38**). *The upper thread must remain in its place IFO the LH horn.* Release the loop you just lifted, dropping it to the back of the horn. As in step 3, you now have a loop of thread encircling each horn and a loose knot sitting behind the LH horn. With your left hand, pull the working thread to the left to tighten the loop against the horn. The work should appear as in **Fig. 41.**

REPEAT STEPS 4 AND 5.

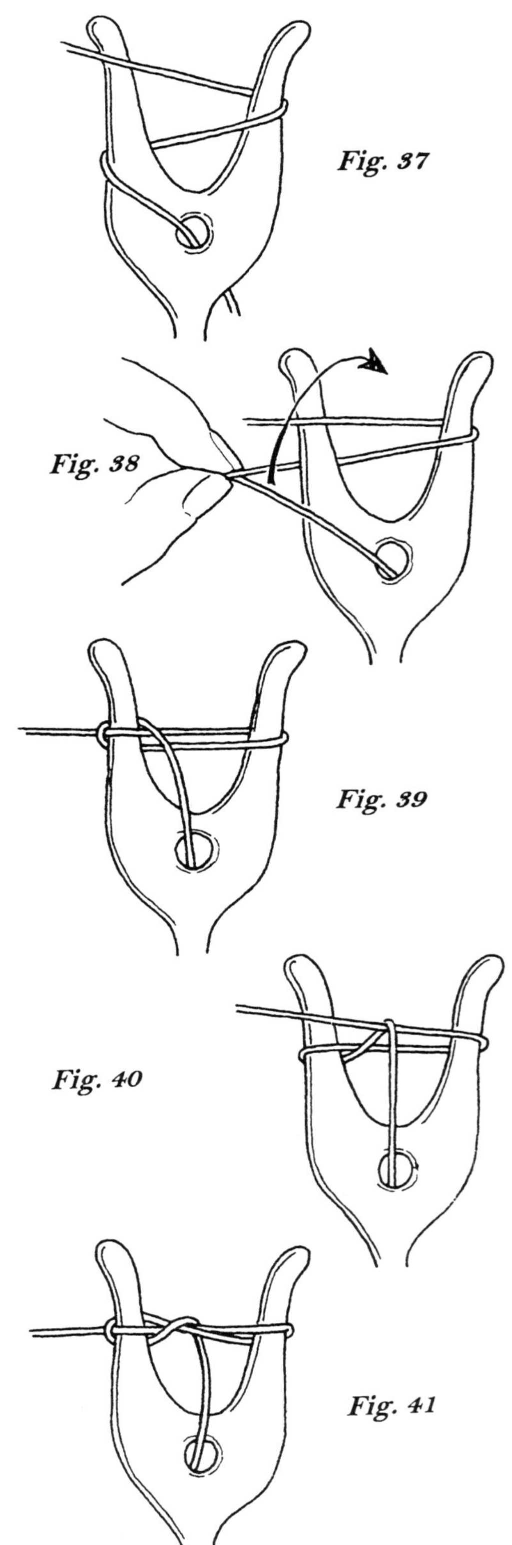

Fig. 37

Fig. 38

Fig. 39

Fig. 40

Fig. 41

Please review the note on tension on page 6, noting that you will be pinching the cord between the thumb and index finger of your right (not left) hand.

ENDING OFF A LEFT-HANDED CORD

1. Turn the lucet as though you were going to start a new stitch, but instead of lifting the loop, cut the thread, leaving about a six-inch tail.

2. Thread the tail down through the loop on the LH horn. Slip the loop off the horn. Pull on the thread until the loop closes.

3. Now thread the tail down through the loop on the RH horn. Slip the loop off the horn. Pull on the thread until the loop closes.

LEFT-HANDED NO-TURN CORD

1. Thread the lucet as in step 1 of the basic Left-Handed cord. Wrap the horns as in step 2 of the basic Left-Handed cord.

2. Slip your left thumb and index finger under the working thread and behind the LH horn. Grasp the lower of the two threads lying between the horns. *Just as you did in step 3 of the basic Left-Handed cord* pull this thread around to the front of the LH horn and lift it up, off, and over the tip of the horn, dropping it behind the LH horn. *The upper thread must remain in its place IFO the LH horn.* You now have a loop of thread encircling the each horn, and a loose knot sitting behind the LH horn. Pull the working thread to the left to tighten the loop against the horn **(Fig 42)**. Continuing to keep tension on the thread, move your hand to the right behind the lucet, stopping just to the right of the RH horn.

3. Bring the thread clockwise around to the front of the RH horn. The new wrap lies above the loop on the RH horn. **Fig. 43.**

4. Grasp the part of the RH loop that lies IFO the RH horn and near the center of the lucet. Move your thumb and index finger to the right, stopping when they are even with the right edge of the RH horn. Lift the loop up past the working thread and off over the tip of the RH horn **(Fig. 44)**. *The working thread must remain in its place on the RH horn.* Drop the loop in the space between the horns. Pull the working thread to the left until the knot is centered between the horns. You may find it helpful to simultaneously pull on the tail with the thumb and index finger of your right hand (your right middle, ring and index fingers are holding the lucet). *Leave the knot a bit loose. You will tighten it in Step 6.*

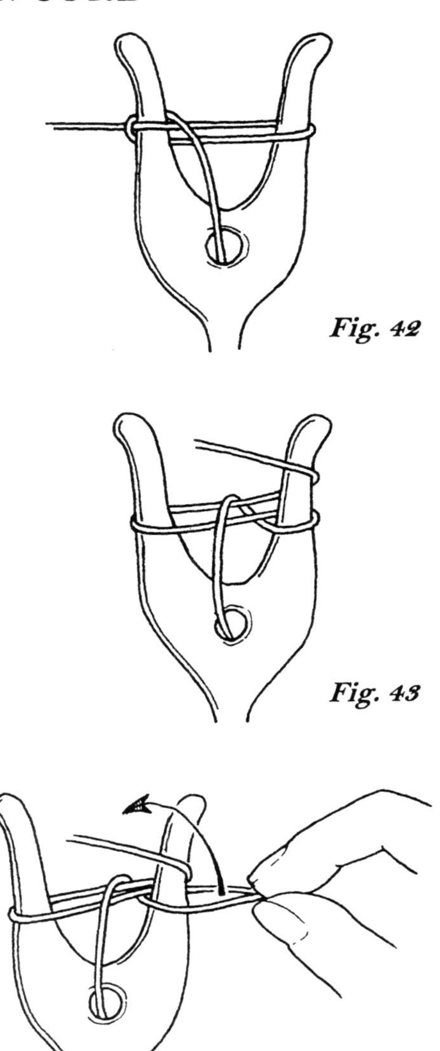

Fig. 42

Fig. 43

Fig. 44

5. Lay the working thread across and IFO the LH horn. The thread should lie above the loop.

6. Grasp the back of the loop on the LH horn. Pull this part of the loop counterclockwise around to the front of the LH horn. When your thumb and index finger have come all the way around to the front of the lucet, tighten the knot from step 4 by pulling on the RH half of the loop. Now lift the loop up, off, and over the tip of horn. Drop the loop behind the LH horn. *The working thread must remain in its place IFO the LH horn.* You now have a loop of thread encircling each horn and a loose knot sitting behind the LH horn. With your left hand, pull the working thread to the left to tighten the loop against the horn. Carry the working thread to the right behind the lucet, stopping just to the right of the RH horn.

N.B.: Steps 5 and 6 make a stitch on the LH horn

7. Bring the thread clockwise around to the front of the RH horn, with the new wrap lying above the loop on the RH horn. **See Fig. 43 again.**

8. Grasp the part of the RH loop that lies IFO the RH horn and near the center of the lucet. Move your thumb and index finger to the right, stopping when they are even with the right edge of the RH horn. If necessary, pull on the back half of the loop to tighten the knot made in step 6. Lift the loop up past the working thread and off over the tip of the RH horn. This is the same motion shown in **Fig. 44.** *The working thread must remain in its place on the RH horn.* When your thumb and index finger reach the center of the space between the horns, release the loop. Pull the working thread to the left until the knot is centered between the horns. For the first few stitches, you may find it helpful to simultaneously pull on the tail with the thumb and index finger of your right hand (your right middle, ring and index fingers are holding the lucet).

N.B.: Steps 7 and 8 make a stitch on the LH horn

REPEAT STEPS 5 THROUGH 8.

When you have enough knots to hang on to, stop pulling on the tail. Instead, hold the cord up near the loops in a pincer grip between the thumb and index finger of your right hand. The lucet will now rest between the palm and remaining fingers of your right hand. This helps keep the tension even.

ENDING A LEFT-HANDED NO-TURN CORD

1. Stop after completing a stitch on the RH horn (Step 8). Cut the thread, leaving about a six-inch tail.
2. Thread the tail down through the loop on the LH horn. Slip the loop off the horn. Pull on the thread until the loop closes.
3. Now thread the tail down through the loop on the RH horn. Slip the loop off the horn. Pull on the thread until the loop closes.

RIGHT AND LEFT TWIST STITCHES FOR LEFT-HANDERS

This is a No-Turn cord.

1. Thread the lucet as in step 1 of the basic Left-Handed cord. Extend your left thumb and wrap the working thread clockwise around it. Coming from below, slip the LH horn of the lucet into the thread loop. Remove your thumb from the loop. The work should appear as in **Fig. 45. You have made a Left Twist Loop on the LH horn.**

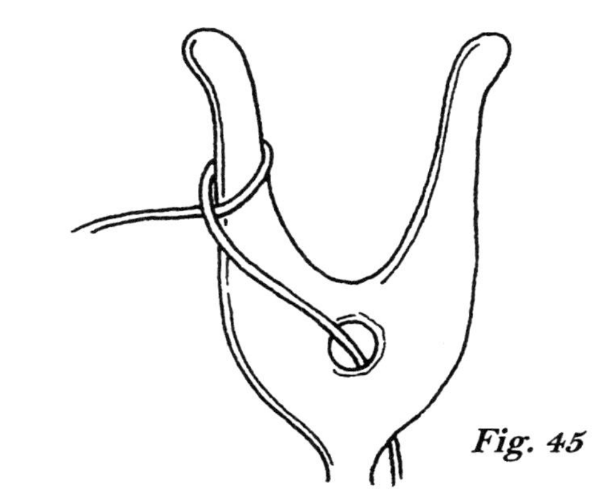

Fig. 45

2. Carry the thread to the right IFO the lucet, stopping when your left hand is just to the right of the RH horn. Extend your right index finger and wrap the working thread counterclockwise around it. Pinch the intersection of the loop between your left thumb and index finger, and withdraw your right index finger from the loop. Now give the loop a half turn away from you by rotating your left hand so that the hand is behind the lucet and the palm faces you. Place the loop on the RH horn of the lucet. Pull the working thread to the left to tighten the loop. The work should now appear as in **Fig. 46. You have made a Right Twist Loop on the RH horn.**

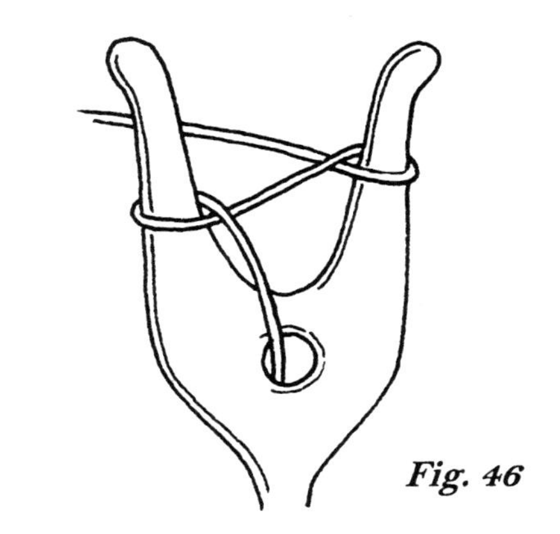

Fig. 46

3. Carry the thread to the left across the back of the lucet. Make a Left Twist Loop on the LH horn. Lift the old loop up, off, and over. Pull the working thread to the right to center the knot, *but do not to tighten it all the way.* **You have made a Left Twist Stitch on the LH horn.**

4. Carry the thread to the right across the back of the lucet. Make a Right Twist Loop on the RH horn. Lift the old loop up, off, and over. Pull the working thread to the left to center the knot, *but do not to tighten it all the way.* **You have made a Right Twist Stitch on the RH horn.**

REPEAT STEPS 3 AND 4.

When lifting the old loop, pull first on the back of the loop to tighten the knot, then on the front to gain enough slack to make the lift.

ENDING OFF THE CORD

When you are finished with this cord, end off as usual for a Left-Handed no-turn cord.

I must confess that I am right handed and that these directions may well not be what a left-handed person would write. They are how I make cord using my left hand. If you are left-handed and come up with a more suitable method, please put it into practice and, more important, teach it to other left-handers! Space limitations do not permit me to translate each set of directions from right- to left-handed. I hope that I have given enough information on basic hand movements for you to be able to work out your own versions of the advanced cords.

TROUBLESHOOTING

CHANGING THREADS

Lucet cord has a high take-up rate. Depending on materials and technique, you may need as much as 10 yards of thread to make one yard of cord. If you are making trim, try to make only as much as you need for a given area (e.g., one length for the neckline and one each for the sleeves). This way you can avoid changing threads altogether. If you are making an eight-yard corset lace with insufficient thread, or the thread breaks, you have two options:

1. Drop a 2-inch tail of the new thread down the back of the cord. Bring the working end of the new thread to the front of the cord just as you would a gimp. Work a few stitches with the old and new threads together. Be careful to pull the knots of the old thread down neatly into the cord. Then drop the old thread to the back of the cord as you would a gimp. Continue working with just the new thread. When you are finished with the cord, clip the ends close and put a drop of nail polish or milliner's glue on them. This makes a lump in the cord, but sometimes it's all you can do.

2. If the cord is all one color, weave the new thread in as a gimp for an inch or so. Then exchange the new thread for the old one, and weave in the old thread as a gimp. Clip the ends and secure as above.

OOPS, OH NO, AND DRAT! (TROUBLESHOOTING)

First, always use working threads that are strong and resistant to abrasion. When you are first learning a particular technique, practice with a strong, smooth thread like pearl cotton. Avoid wool until you know the pattern and are less likely to make mistakes. When you do use wool, choose a yarn that is difficult to break by hand. Avoid mohair and angora altogether. At the other extreme, a hard, tightly spun thread can snarl and break. Some rayons can stand up to abrasion from the horns; others cannot.

Some workers find it helpful to mark one side of the lucet with a tiny dot of colored nail polish or permanent marker.

If you have put the lucet down and now are uncertain which horn should get the next stitch, look carefully at the loops on both horns. One loop will probably appear to sit slightly higher on its horn than the other. This is the last loop made. You should make the next stitch on the *other* horn. Furthermore, the working thread should come in a straight line out of the last loop and run across to the horn that should get the next stitch (**Fig. 47**).

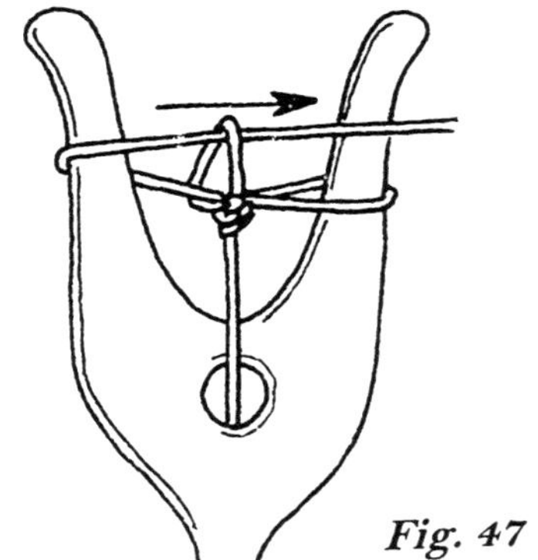

Fig. 47

If you thread breaks close to the horn, or if you make a mistake in a patterned cord, you will need to unpick.

1. Carefully pull on the front and then the back of the last knot. One side will tighten the knot and one will loosen it. A crochet hook or blunt needle is helpful for this operation. Once you have found which way to pull to loosen the knot, keep at it until you have a loop big enough to slip back onto the horn.

2. Place the loosened loop back on the horn, being careful to preserve the twist if you have been working twisted stitches. Unwrap the working thread from around the horn.

3. Repeat the process of loosening, replacing the loop on the horn, and unwrapping until you have corrected the mistake or have regained enough thread to make a secure join.

APPLICATIONS

Re-enactors, members of the Society for Creative Anachronism, and participants in Renaissance Faires will all find the lucet a valuable tool. Idle moments in camp can be spent making lacing cord and trim to match your costume, from your own design, with a tool that fits easily into your workbasket, pouch, pocket, or sleeve.

The basic single-thread cord makes excellent lacing cord. It is strong, fairly inelastic and serves well when tight lacing is required. The cord also wears well when used for drawstrings. There is evidence that the cord found ecclesiastical usage, probably as trim. Lucet cord is well suited to the Medieval and Renaissance practice of couching cord over the edges of appliqués. Unbeaded cords, worked in silk and tipped with an aiglette, make excellent points for costume use. Worked with several strands of silk held together and finished with a tassel, any of the unbeaded, square cords make a nice mantle lace. Interestingly enough, the mantle lace from Cosimo de' Medici's grave clothes (1574) shows the square profile, V-shaped stitches, and characteristic longitudinal indentations of the Cord With Two Working Threads, Method I.

Cordwork or braiding was a popular form of surface embroidery in both the Renaissance and Victorian eras. It is difficult, to say the least, to find silk braid of the fineness used in Victorian braiding patterns. Cords worked in either buttonhole silk or pearl cotton (size 8 or 12) make a good substitute. Square cords can be couched, while flatter ones should be sewn with tiny running stitches down the middle. Here are two cordwork patterns for you to try. The first is from the February, 1873 issue of Peterson's Magazine. This is a simple pattern, requiring only one line of cord. *Enlarge as appropriate for your particular use.*

The second pattern is taken from the cordwork on the *ropa* of Emilia di Spilimbergo, c. 1560. This pattern requires two separate cords. As is typical of Renaissance cordwork, the lines pass alternately over and under each other. Don't panic; just find a blunt needle with an eye large enough to take the cord easily. Start with the line marked "1". Couch this line down all the way to the end of your pattern. Now thread the blunt needle with the second length of cord. Lay the cord along the pattern line marked "2". Couch until you get to the circled intersection. Slip the needle under the first cord and carefully pull the second cord through. Do this every time you need to pass the second cord under the first. At the end of the pattern, take the tails of both cords through to the back of the fabric and tack them down along the pattern lines. *Enlarge as appropriate for your particular use.*

Twisted stitch cords in size 8 pearl cotton can make delicate Battenburg lace pieces. With a photocopier, reduce the Battenburg pattern until the width of the tape lines matches the width of your braid. Then baste the braid down and proceed as usual. If your design has a lot of short pieces and joins, you may want to measure the pieces and make corresponding lengths of cord, rather than making one long strip and then cutting it. Cords with beaded gimps make particularly sumptuous Battenburg miniatures.

Puncetto and Armenian lace are usually worked directly on a hem. If the fabric wears out, or the article has An Unfortunate Incident, the lace has to be cut away. If made on a header of twisted stitch cord, the finished lace can be whipped down and removed as desired. Work the first line of stitches into one edge of the cord.

Edwardian tatters and crocheters incorporated a variety of novelty braids into their work. One of these braids had picots on both edges, similar to lucet picots. Another, called coronation cord, had thick-and-thin sections along the length of the cord. This can be approximated in lucet cord with size 5 pearl cotton. Alternate four plain and twelve twisted stitches. Make an even number of repeats. Form the cord into a zigzag by bending it at the thin spots.

BIBLIOGRAPHY

Andere, Mary. *Old Needlework Boxes And Tools: Their Story And How To Collect Them.* New York: Drake Publ., 1970.

Epstein, Kathleen. *German Renaissance Patterns For Embroidery.* Austin: Curious Works Press, 1994.

Fischer, Birthe Karin. *Band og redskab.* Viborg: Sesam, 1978.

Goodnow, Kendra, and Hilliger, Albert C. *Lucette Cord Made Easy.* 1998.

Groves, Sylvia. *The History Of Needlework Tools And Accessories.* Feltham, Middlesex: The Hamlyn Publishing Group, 1966.

Proctor, Molly. *Needlework Tools and Accessories: A Collector's Guide.* London: B. T. Batsford, Ltd., 1990.

Rogers, Gay Ann. *An Illustrated History Of Needlework Tools.* London: John Murray Pubs., 1983.

Wilkinson, Dorothy S. "Needlework Tools: The Lucet." *Needle Arts,* June 1997, pp. 16-17.

SOURCE

LACIS, 2982 Adeline Street, Berkeley, California 94703, tel: 510-843-7178; fax: 510-843-5018; E-mail: staff@lacis.com; Web: www.lacis.com

Lucets, kumi himo bobbins, threads, metallic braids, ribbons, beads and classes.